# Innovations of Green Process Engineering for Sustainable Energy and Environment



Topical Conference at the 2012 AIChE Annual Meeting

## Pittsburgh, Pennsylvania, USA
## 28 October - 2 November 2012

ISBN: 978-1-62276-751-9

**Printed from e-media with permission by:**

Curran Associates, Inc.
57 Morehouse Lane
Red Hook, NY 12571

**Some format issues inherent in the e-media version may also appear in this print version.**



Printed by Curran Associates, Inc. (2013)

For permission requests, please contact AIChE
at the address below.

AIChE
3 Park Avenue
New York, NY 10016-5991

Phone: (203) 702-7660
Fax: (203) 775-5177

www.aiche.org

**Additional copies of this publication are available from:**

Curran Associates, Inc.
57 Morehouse Lane
Red Hook, NY 12571 USA
Phone: 845-758-0400
Fax: 845-758-2634
Email: curran@proceedings.com
Web: www.proceedings.com

# TABLE OF CONTENTS

# (44b) Conceptual Net Energy Output for Biofuel Production from Lignocellulosic Biomass Through Biorefining: A Review

**J.Y. Zhu**, USDA Forest Service, Forest Products Laboratory, Madison, WI

**Conceptual net energy output for biofuel production from lignocellulosic biomass through biorefining: A review**

J.Y. Zhu[1,2]

[1] USDA Forest Service, Forest Products Laboratory, Madison, WI, USA

[2] Dept. Biological Systems Engineering, University of Wisconsin-Madison, WI, USA

**Abstract**

The concept of producing lignocellulosic biofuel, bioproducts, and chemicals through a biorefinery using lignocelluloses has been around for 70 years or more. The recent interest in this old concept arises from the promises of mitigating climate change by substituting some biomass energy for petroleum or other fossil-fuel energy; achieving national energy independence and sustainable economic development; and advances in biotechnology in the past 2-3 decades. Geopolitical and national security reasons have greatly contributed to the need for seeking alternative energy, especially using domestic renewable and sustainable sources.

However, there is a lack of comprehensive information in the retrievable literature on pilot scale process and energy data using promising process technologies and commercially scalable and available capital equipment for lignocellulosic biomass biorefining, despite several pilot scale and demonstration plants are under construction. Unfortunately, bioenergy production is not sustainable without net energy production. Recognizing the heated debate on energy production from lignocellulosic biomass in the scientific community and most arguments were not based on actual laboratory data for biorefinery operation, this presentation is intended to provide an objective view on energy production from lignocellulosic biomass. The process data from approximately a dozen studies that represent state-of-the-art in cellulosic biofuel production concepts, along with literature energy input data for agriculture operations, were analyzed to provide estimates of net energy production. Several scenarios were created for proper allocation energy input to biomass and major agriculture co-product, such as corn or lumber.

For biorefinery operations, increasing solids loading in pretreatment to at least 25% is critical to reducing energy input in a biorefinery. Post thermo-chemical pretreatment size reduction approach should be adopted for energy efficient woody biomass processing. When appropriate pretreatment technologies are used, woody biomass can be processed as efficiently as herbaceous biomass and agricultural residues. Net energy output for cellulosic ethanol was estimated to range approximately from -500 to 2000 MJ/ton biomass (HHV base); indicating that the energy input/output ratio is approximately 1:1 for cellulosic ethanol. However, net energy can reach approximately 4000-7000 MJ/ton of biomass when energy from lignin is included.

# (44c) The R&D of Cellulosic Ethanol in China

**Xinshu Zhuang**, Bio-chem conversion lab, Guangzhou institute of energy conversion,Chinese Academy of Sciences, Guangzhou, WI, China

As the biggest developing country, China needs fuel ethanol to meet its demands of its social and economy development. Basing on the "Mid-long term plan for renewable energy" established in 2007, the annual consumption of fuel ethanol will reach 10 million tons in 2020. Cellulosic ethanol, one of the most important second generation bio-fuel, gains more and more attentions for its substitution of the transportation liquid, lessen the air pollution by the vehicles and promote the rural area economy. In China, more than five demonstration projects have been set up since 2008. In this paper, the main cellulosic ethanol processes applied in China are illustrated, simultaneously the R&D of a green and high efficient process pretreated with hot liquid water is presented.

# (53a) Real-Time Operation and Control of Integrated Ultrafiltration-Reverse Osmosis Membrane Desalination System

**Larry Gao**[1], Han Gu[2], Anditya Rahardianto[2], Panagiotis D. Christofides[3] and Yoram Cohen[4], (1)Chemical & Biomolecular Engineering, UCLA, Los Angeles, CA, (2)Chemical & Biomolecular Engineering, University of California, Los Angeles, Los Angeles, CA, (3)Department of Chemical and Biomolecular Engineering, University of California, Los Angeles, Los Angeles, CA, (4)Chemical and Biomolecular Engineering, University of California, Los Angeles, Los Angeles, CA

Reverse osmosis (RO) membrane desalination has emerged as one of the leading methods for water desalination due to the low cost and energy efficiency of the process. Recent advances in model-based control and energy optimization for RO desalination [1-3] have made it possible to develop self-adaptive operational strategies for RO membrane desalination systems. The goal of self-adaptive operation is to enable the plant to autonomously respond to changes in feed water salinity and fouling propensity. In order to combat membrane fouling, RO feed water pre-treatment is crucial for effective RO desalination. In recent years, microfiltration (MF) and ultrafiltration (UF) systems have been increasingly explored for effective feed-filtration. While the concept of UF/RO integration is appealing, there are significant technical challenges in situations where plant footprint (e.g., for shipboard desalination) is at a premium, the use of cleaning chemicals must be kept to a minimum, and feed water quality often fluctuates. To demonstrate the feasibility of an integrated UF/RO system and the operational benefits of advanced process control, the UCLA team developed a compact second generation RO (CoM2RO) desalination system integrated with MF-UF feed filtration. The CoM2RO implements two unique controllers: a novel RO model-based controller capable of simultaneously self-adjusting the feed flow rate and trans-membrane pressure in order to maintain the desired water productivity and quality, and a self-adaptive UF operational strategy that varies backwashing intensity depending on detected fouling levels. These controllers enable integration of the two systems such that the UF can directly utilize the RO concentrate for backwashing and minimizes RO energy consumption, while the RO system adapts to any changes in pressure, flow rate, or salinity that result from the cyclic nature of the UF backwashing. Real-time sensors data along with embedded computing capability allow on-line membrane element characterization and overall system performance assessment (with appropriate data normalization) that are used for rapid control and optimization. A field evaluation of the present advanced RO/UF system control was carried out for seawater desalination. Experiments demonstrated that the pre-treatment operation significantly improved the RO feed water conditions (Turbidity, florescence reading, particle size) which the RO controller successfully adapted to while maintaining constant RO permeate production and quality. Real-time sensor data allowed for on-line system characterization of fouling levels in the UF system, thereby enabling adjustment of operating conditions as well as fault detection and significantly lengthening the operational duration of the system. Results of the study demonstrated that the real-time information regarding membrane characterization enabled fault detection and rapid evaluation of the impact of varying fouling and feed conditions on the performance of both UF and RO systems.

1. Zhu, A., P. D. Christofides and Y. Cohen, "Effect of Thermodynamic Restriction on Energy Cost Optimization of RO Membrane Water Desalination," Industrial and Engineering Chemistry Research, 48, 6010-6021 (2009).
2. Bartman, A.R., A.H. Zhu, P.D. Christofides, Y. Cohen, "Minimizing Energy Consumption in Reverse Osmosis Membrane Desalination using Optimization-Based Control," J. Process Control, 20, 1261-1269 (2010).
3. Zhu, A., P. D. Christofides, and Y. Cohen, "Energy Consumption Optimization of Reverse Osmosis Membrane Water Desalination Subject to Feed Salinity Fluctuation," Ind. Eng. Chem Res, 48, 9581–9589(2009).

# (53b) Optimal Plant Operation of Industrial Reverse Osmosis Desalination

**Mingheng Li**, Department of Chemical and Materials Engineering, California State Polytechnic University, Pomona, CA and Brian Noh, Inland Empire Utility Agency, Chino, CA

A comprehensive first-principles based mathematical model, which explicitly accounts for membrane hydraulic permeability, pump characteristic curves and retentate pressure drop, is developed to describe the reverse osmosis desalination process. Parameters used in the model are derived from data collected in an industrial desalination plant. A model-based optimization framework is then developed to reduce pump energy consumption while maintaining the same permeate rate. The model predicted results match well with plant trials, and a 10% reduction in energy consumption is demonstrated.

References

[1] Li, M. "Optimization of Multitrain Brackish Water Reverse Osmosis (BWRO) Desalination," Industrial Engineering Chemistry Research, 51, 3732-3739, 2012.
[2] Li, M. "Optimal Plant Operation of Brackish Water Reverse Osmosis (BWRO) Desalination," Desalination, 293, 61-68, 2012.
[3] Li, M. "Reducing Specific Energy Consumption in Reverse Osmosis (RO) Water Desalination: An Analysis from First Principles," Desalination, 276, 128-135, 2011.
[4] Li, M. "Minimization of Energy in Reverse Osmosis Water Desalination using Constrained Nonlinear Optimization," Industrial and Engineering Chemistry Research, 49, 1822-1831, 2010.

# (53c) Novel Cellulose Esters for Forward Osmosis Membranes

**Rui Chin Ong**[1], Jincai Su[1], Bradley J. Helmer[2], Jos S. de Wit[2] and Tai-Shung Chung[3], (1)National University of Singapore, Singapore, Singapore, (2)Eastman Chemical Company, Kingsport, TN, (3)Department of Chemical and Biomolecular Engineering, National University of Singapore, Singapore, Singapore

In view of the tremendous efforts for the invention of materials and techniques for forward osmosis (FO) membrane fabrication, in this work, we focused our study in the use of novel cellulose ester materials for FO membrane fabrication. A wide range of cellulose esters with different degree of distributions (DS) of different functional groups which include hydroxyl (OH), acetyl (Ac), propionyl (Pr) and butyryl (Bu) groups were specially synthesized for this study. The effects of DS of functional groups of the cellulose esters have on the resultant membranes' morphology and performance were carefully studied. The resultant membranes have distinguished morphology and these cellulose esters exhibited distinguished membrane performance under forward osmosis process due to their differences in the functional group distribution. Fundamental properties such as water permeability, water solubility, salt partition coefficient and salt permeability of each cellulose ester were characterized to determine the effects of functional group distributions on the membrane performance. This study showed that desired membrane performance can be achieved by altering the DS of various functional groups for cellulose esters.

# (53d) Harvesting of Pure Water From Air and Combustion Exhaust Gas by Membrane Pervaporation

**Wei Liu**, Pacific Northwest National Laboratory, Richland, WA

When there is scarcity in natural water resources, supply of pure, potable water becomes a problem for sustainable living and industrial development. On the other hand, presence of water in air is ubiquitous and water is also produced from combustion of biomass and hydrocarbon fuels. This presentation presents a brief review of various water recovery technologies reported in the literature and introduces a new membrane technology for production of water from humid air and/or combustion flue gas via a pervaporation process. In the pervaporation process, as a water-containing gas stream flows over a membrane surface, water vapor is selectively pulled out of the mixture into the another side of the membrane and is compressed to above its dew point by a vacuum pump so that the water vapor is condensed at ambient temperature and collected as liquid water. Important performance requirements of the membrane are 1) high water permeance, 2) high selectivity of water vapor over air, 3) robustness of the membrane, and 4) low cost. The thin flat sheet molecular sieve membrane recently developed in our research group will be presented to address these critical requirements for a practical membrane water production system. The new membrane provides very high water vapor permeance and a large window of operation conditions, such as temperature and water vapor content. Due to the water molecular specificity of the zeolite membrane material, the membrane pervaporation produces pure water in one-step without need of further purification.

# (53e) Design and Development of Hollow Fiber Membrane for Desalination and Salt Recovery Via Integrated Membrane Distillation-Crystallization

**Felinia Edwie**, Chemical and Biomolecular Engineering, National University of Singapore, Singapore, Singapore and Tai-Shung Chung, Department of Chemical & Biomolecular Engineering, National University of Singapore, Singapore, Singapore

Over the past few years, the water research worldwide has carried a similar theme of water crisis with the continuous increment of world population and limited availability of fresh water resources were identified as the major causes. In this regard, desalination of saline water has great potencies to suffice the world's increasing demand of fresh water considering the high percentage of saline water on earth. Among other membrane-based desalination process, reverse osmosis (RO) technology has been proven reliable and becomes the most preferred option for desalination. Nevertheless, the large scale application of RO is often questioned for the brine disposal issue and its unfavorable impacts on the environment. These issues have urged the development and innovation of more sustainable desalination processes with the ultimate aim to maximize the water recovery and the process efficiency. For these reasons, various hybrid desalination processes have been proposed, for instance membrane distillation (MD) based hybrid desalination technology. MD process is an integration of membrane-based and thermal-based desalination processes in which a porous hydrophobic membrane is employed as a thermal insulator and physical barrier between the hot feed where water evaporation occurs and the cold distillate where the diffused water vapor is condensed. Therefore, the permeation flux in MD process is less dependent on the solution concentration as the transport of water vapor is driven by the vapor pressure gradient across the membrane. This allows MD operation to concentrate the saline solution up to saturation level enabling the subsequent recovery of the salt by means of the crystallizer. Therefore, by integrating membrane distillation with crystallizer (MDC), not only the valuable compounds can be harvested but the issue of brine disposal can be also mitigated.

As compared to MD operation with a low concentration of saline solution, membranes operated with a high concentration are more prone towards salt scaling facilitated membrane wetting which results in flux decay and deterioration of separation performance. Therefore, in this study we investigated the performance of various hollow fiber membrane configurations under MDC operation with a highly concentrated sodium chloride (NaCl) aqueous solution. The membranes were developed in our laboratory which include single- and dual-layers membrane configurations along with combination of hydrophobic-hydrophilic characteristics. Heat and mass transfers along membrane module were modeled to determine the change of membrane permeability with increasing salt concentration. Upon MDC operation, the membrane performances were retested to define the recovery of membrane permeability. It was found that the morphology underneath membrane surface played important role in mitigating scaling facilitated membrane wetting and membrane with cellular structure exhibited a better wetting resistance as compared to globular structure. Moreover, the performance of membrane with cellular structure was maintained even at the widest trans-membrane temperature gap at which the scaling propensity is the highest. Lastly, NaCl was crystallized and recovered from the supersaturated MDC retentate solution. The effects of crystallizer cooling rate on the resultant salt crystal nucleation and growth were investigated.

# (53f) Prevention of Environmental Pollution with Advanced High Capacity Water Storage

**Lev Reznikov**, Engineering, Global Fabrication, Inc., DuBois, PA and Luke Sicard II, Engineering, Global Fabrication Inc., DuBois, PA

Oil and gas companies throughout the world are doing their part to protect nature and the environment, and consumers are paying closer attention than ever to make sure they do.

Damage to local wildlife can bring negative publicity, regulatory fines, and other long-term expenses to your operation.

Global Fabrication, Inc. designed and introduced the most efficient on the market waste water storage tanks that preventing any considerable leakages to the surroundings.

Full assembly of the high capacity tanks takes much less time that the competitive storages.

Novel technological concept provides not only significant cost reduction technology, but also practically full prevention of leakages.

It is our promise to safe life of all kinds of birds and animals!

A system of weaved netting around your Big Holding water storage tanks can easily prevent small and large creatures from harming themselves or damaging your equipment. The result is a commonsense system that's as good for business as it is local wildlife.

A system of weaved netting around your Big Holding water storage tanks can easily prevent small and large creatures from harming themselves or damaging your equipment. The result is a commonsense system that's as good for business as it is local wildlife.

# (148a) Novel Materials for Carbon Capture, Utilization and Storage (CCUS)

**Ah-Hyung Alissa Park**, Earth and Environmental Engineering & Chemical Engineering, Columbia University, New York, NY

Historically, the atmospheric concentration of $CO_2$ fluctuated naturally on the timescales of ice ages. Concerns, however, stem from the recent dramatic increase in $CO_2$ concentration, which coincides with global industrial development. This rise is mainly due to the high use of fossil fuels. In order to meet the ever-increasing global energy demands while stabilizing the $CO_2$ level in the atmosphere, it is widely believed that current carbon emissions must be reduced by at least a factor of three. The containment of $CO_2$ involves three operations: separation, transportation, and storage. Until now, these technologies have been developed independent of one another, which has resulted in complex and economically challenging large-scale designs. The future direction of carbon management technologies now focuses on the integration of $CO_2$ capture and storage schemes as well as $CO_2$ utilization. In this seminar, two novel carbon capture, utilization and storage (CCUS) technologies will be introduced. $CO_2$ capture fluids based on the Nanoparticle Organic Hybrid Materials (NOHMs) are currently developed and their absorption isotherms are characterized as a function of $CO_2$ partial pressure and temperature (i.e., combustion and gasification conditions). NOHMs are a new class of organic-inorganic hybrids that consist of a hard nanoparticle core functionalized with a molecular organic (sometimes polymeric) corona that possesses high degree of tunability. NOHMs are non-volatile and stable over a very wide temperature range, which make them interesting materials for various energy and environmental applications. Once captured, $CO_2$ needs to be stored for permanent disposal. The geological storage of carbon dioxide has been considered to be the most economical method of carbon sequestration, while mineral carbonation is a relatively new and less explored method of sequestering $CO_2$. The advantage of carbon mineral sequestration is that it is the most permanent and safe method of carbon storage, since the gaseous $CO_2$ is fixed into a solid matrix of Mg-bearing minerals (e.g., serpentine) forming a thermodynamically stable solid product. These carbon sequestration technologies can be integrated into the existing or new energy conversion systems in order to achieve their overall sustainability.

# (148b) On the Selection of Ionic Liquids for the Conversion of Cellulosic Biomass Feedstock

**Sameer Parvathikar**, Andrew R. Tadd and Johannes W. Schwank, Chemical Engineering, University of Michigan, Ann Arbor, MI

The use of ionic liquids (ILs) as solvents represents a promising route for the conversion of cellulosic biomass feedstock. 1,3-dialkylimidazolium ILs in particular, have become a popular and thoroughly-studied reaction media for this. Reactions such as the dissolution and hydrolysis of cellulose, the dehydration of glucose, etc. have been demonstrated. While these ILs have proven to be quite capable of processing these large biomolecules, their widespread adoption has been limited due to their toxicity and inhibitive costs.

It is clear that there is a need for less expensive, and environmentally benign alternatives. The objective of this work is to develop a deeper understanding of the interaction between the solute (biomolecules) and the solvent, which is necessary to effectively choose these alternative solvents. A number of solvent properties like hydrogen bond formation, acidity, etc., will be studied for both imidazolium-based ILs and lower-cost alternatives. Using probe reactions like the isomerization and dehydration of sugars, property-performance relationships will be established. The effect of using different catalysts to affect the reaction environment and thence product yields will also be examined. This work will provide insights into rationally selecting ILs to perform targeted chemistries.

# (148c) Production of Hydroxymethylfurfural From Sugars or Renewable Biomass Feedstock As a Potential Platform Intermediate for Bio-Plastics

**Wei Liu**, Pacific Northwest National Laboratory, Richland, WA

There has been world-wide interest in making plastics out of renewable biomass feedstock for recent years. Hydroxymethylfurfural (HMF) is viewed as an attractive alternate to terephthalic acid (TPA) for production of polyesters (PET) and polyamides. An economic production process needs to be developed to make HMF at a cost competitive with current petroleum-derived feedstock. Conversion of sugars into HMF has been studied in numerous publications. In this presentation, a brief overview of those literature reports pertinent to a practical process development will be presented. A complete ionic liquid-based catalytic process recently developed at PNNL will be discussed. This new process comprises integration of ionic liquid catalytic conversion with adsorption separation. The ionic liquid catalytic reaction is conducted under moderate reaction conditions and provides both high reactivity and selectivity, which is practical for scale-up. The main challenge lied in separation of reaction mixtures. Thus, novel membrane-coated zeolite adsorbent particles are invented, and the adsorption process is demonstrated for isolation of 99% pure HMF product and complete recovery of the ionic liquid and un-converted species. The new process enables nearly 100% conversion of fresh fructose feed into HMF product in the recycled ionic liquid. A simplified process flow diagram is conceived based on these innovations, and the key equipment such as reactor and adsorbent bed is sized for a plant of 200,000 ton/year of fructose processing capacity. The proposed HMF production process is much simpler than the current paraxylene (PX) manufacturing process from petroleum oil, which suggests substantial reduction to the capital cost and energy consumption is possible. At the equivalent value to PX on the molar basis, there can be a large gross margin for HMF production from fructose and/or sugars.

# (148d) Large-Scale Energy Storage: New Generation Vanadium Mixed Acid Redox Flow Battery

**Liyu Li**, UniEnergy Technologies, LLC, Mukilteo, WA

Electrical energy storage is considered as a key enabler to significant penetration of intermittent renewable power from solar and wind, as well as implementation of the Smart Grid. Among the most promising technologies is the all-vanadium redox flow battery (VRFB), which is capable of storing electricity on a MW/MWh scale in the form of two pairs of reduced and oxidized vanadium species ($V^{2+}/V^{3+}$ and $V^{4+}/V^{5+}$) dissolved in two separate liquid electrolytes, for a discharge duration of hours or even longer. VRFBs offer the flexibility of separate design of power (kW) and energy (kWh): power depends on stack design, while energy depends on the volume and vanadium concentrations in electrolytes. With electrodes free of repetitive mechanical or structural stresses, VFRBs potentially deliver virtually unlimited cycle life. In addition, VRFBs are inherently safe as flowing electrolytes carry away the heat generated during operation. Other advantages include quick response, deep discharge ability, capability to withstand a fluctuating power supply, tolerance to over-charge and over-discharge, and fast charge and discharge. Substantial advancement has been made in VRFBs using the traditional vanadium sulfate electrolytes in the past years. However this technology cannot meet the economic and performance requirements for broad market penetration. While there is much room to improve the economy and performance by engineering and system integration, the current technologies are inherently limited by the current electrolyte chemistries. To prevent irreversible vanadium oxide precipitation, active control of operating temperatures in the range 10-35$^{o}$C is essential, as is vanadium concentrations (i.e. energy capacity) $\leq$ 1.7M (1.5M in practical systems). Many impurities have to be controlled down to ppm levels to maintain the stability of the electrolytes.

Recently, new vanadium redox chemistry with $SO_4^{2-}/Cl^-$ mixed supporting electrolytes was invented.[1] To stabilize the vanadium electrolytes, chloride was added into the aqueous sulfate electrolyte solution to form new complexes of vanadium species. The new complexes were proved to be much more stable over the current vanadium sulfate chemistries, leading to: i) an extension of operation temperatures to a range of -5 to 60$^{o}$C, potentially eliminating the need for external heat management; ii) an increase in concentrations of vanadium species ($V^{5+}$, $V^{4+}$, $V^{3+}$, $V^{2+}$) to 2.5M, representing an energy capacity increase of 70% when compared to the all-vanadium sulfate chemistry; iii) achievement of nearly 90% energy efficiency at 50mA.cm$^{-2}$, with negligible degradation over a few hundred cycles. In this presentation, the structure of the vanadium species in the new electrolyte solutions and the performance and large-scale implementation potential of this advanced VRFB system will be discussed.

*Acknowledgements*

Part of the work was performed at Pacific Northwest National Laboratory, and supported by US Department of Energy.

*References*

1. Li*, Kim, Wang, Vijayakumar, Nie, Chen, Zhang, Xia, Hu, Graff, Liu, Yang*, *Advanced Energy Materials*, 1, 394~400, 2011.8th International Vanadium Symposium: Chemistry, Biological Chemistry, & Toxicology (vanadium8.org)

# (148e) High Surface Area Transparent Conducting Oxide Electrodes and Their Application to Energy Technologies Based On Electrochemical Processes

**Arnold J. Forman**, Zhebo Chen and Thomas F. Jaramillo, Chemical Engineering, Stanford University, Stanford, CA

The ability to control the interfacial area and light trapping characteristics of structured electrodes enables new levels of performance in modern optoelectronic devices. Textured, high surface area electrodes (HSEs) combine large interfacial areas and increased utilization of active material with a light trapping morphology – key parameters for boosting device performance. Transparent conductive oxide (TCO) HSEs with tunable pore sizes on the nanometer to micron scale facilitate diffusion through the structure and serve as a broadly applicable platform for functionalization in many fields including electrochemical sensors, catalysis, (super)capacitors, batteries, photovoltaics and photoelectrochemistry. In this talk we describe the development of optically transparent, electrochemically stable, physically robust HSEs with tunable roughness factors (RF) from 1 through >100. We present a bottom-up approach to future device design and a low-cost, scalable synthetic route for production of transparent HSEs of several TCO compositions. Lastly, we demonstrate the utility of these HSEs by incorporating thin films and particles of electrochemically active materials for enhanced performance. Examples will include adopting the HSE platform for improved supercapacitors and solar photoelectrochemical water-splitting, among others.

# (148f) Novel Wave-Shaped MnO2 Cathode Supported by a Partially Porous Nickel Layer for Fuel Cell/Battery

Mike Musil, Institute of Industrial Science, The University of Tokyo, Tokyo, 153-8505, Japan and Atsushi Tsutsumi, Institute of Industrial Science, The University of Tokyo, Tokyo, Japan

Increasing global demand for green energy has boosted the development of various means of energy generation and storage. Both fuel cells and batteries show big potential for an increasing amount of applications, however they still suffer from various drawbacks. Among others, fuel cells possess a low power density, which can mainly be ascribed to the reaction that occurs at a three-phase interface between electrode, electrolyte and cathodic and anodic reactants. Batteries, on the other, hand lack high energy density, whose reason lies in the limited amount of active material in the electrodes.

To eliminate each one's drawback, a so called Fuel Cell/Battery (FCB) has been suggested in previous works. As the name implies, it can work both as a fuel cell and battery. When working as latter, the best results were obtained when coating $MnO_2$ via electrochemical deposition on carbon fibers. However, oxygen charging – crucial when running in fuel cell mode – was fairly poor. Therefore, a novel cathode structure is presented in this work.

On the one hand, a high surface area to enhance the oxygen charging rate is obtained via a wave-shaped structure in the micrometer scale. Thanks to the obtained surface increase, a reaction rate improvement by up to one order of magnitude is expected compared to a flat electrode. To tackle the problem of having to charge electrically insulating $MnO_2$ via oxygen while enabling electron collection, a novel, tunnel-looking structure has been developed. Via template method, a several micrometer thick Nickel electrode with one micrometer long pillars has been obtained through electrochemical deposition. On top of the pillars, $MnO_2$ is deposited, creating a triple layer consisting of $MnO_2$, porous Nickel, and bulk Nickel.

# (220a) Multi-Objective Optimization Applied to Energy Efficient-Buildings

**Ekaterina Antipova**, Department of Chemical Engineering, University Rovira i Virgili, Tarragona, Spain

Multi-objective optimization applied to energy efficient-buildings.

ABSTRACT

E. Antipova[a] , D. Boer[b], L. Cabeza[c], G. Guillén-Gosálbez[a], L. Jiménez[a]

a. Department of Chemical Engineering, University Rovira i Virgili

b. Department of Mechanical Engineering, University Rovira i Virgili

c. Edifici CREA, Universitat de Lleida, Pere de Cabrera s/n, 25001-Lleida, Spain

The growth of population implies the growth in construction speed and housing cost. Affordable housing is one of the global problems in the world. Though the term "affordable" implies some compliance with the environmental regulations, the main target of the construction is to make the housing cheaper. However, cost reduction is not the unique criterion that should be taken into account when projecting a new building. Buildings are considered to be responsible for up to 50% of the total $CO_2$ emissions worldwide (if we consider both direct and indirect impacts). Hence, this is a sector with large potential for environmental improvements that demands adequate measures for mitigating global warming. The problem of energy efficient buildings is equally relevant for cold and hot climates. In the former, it is necessary to insulate the building properly to keep it warm in winter months, while in the latter, we need to cool it down in summer and to maintain a comfortable temperature in winters as the insulation of the buildings in these regions is usually poor because of the large-scale tendency to build low-cost housing.

When we talk about direct contribution of the construction to the environmental impact, we refer to the use of energy during operation of buildings (heating, cooling, ventilation, lightening and etc.), while indirect contribution accounts for the construction and maintenance of the buildings (embodied energy). It is important to make preliminary estimations to balance both constituents: operation and maintenance, as sometimes reduction in the operation impact causes the growth of the impact related to construction. This might happen, for instance, when we decrease the use of fossil fuels to heat the spaces by increasing the insulation of the building, thereby offsetting the savings in operation by the additional impact due to the production of isolating materials.

In this work we propose a systematic method for the retrofit of buildings considering economic and environmental aspects. The design is posed mathematically as a multiobjective mixed-integer nonlinear programming model (MINLP) that considers the simultaneous minimization of cost and environmental impact given a certain house to be retrofitted. Our model accounts for a set of retrofit measures to make a building more energy efficient, such as incorporation of alternative types of windows, variation of insulation of walls and roof, and employment of different types of solar collectors for hot water generation, among some others. The environmental performance is quantified following life cycle assessment (LCA) principles. A case study based on existing facilities of cubicles located in Lleida (Spain) is used to illustrate the capabilities of the approach. By applying the epsilon constraint method, a set of Pareto solutions representing the optimal trade-off between both conflicting objectives is generated. The systematic tool presented herein is intended to guide decision-makers towards the achievement of economic and environmental viability of the construction processes.

# (220b) Thin Flat Sheet Molecular Sieve Membranes for Air Drying and Energy Recovery of Building Ventilation

**Wei Liu**, Pacific Northwest National Laboratory, Richland, WA

Dehumidification and/or humidity control of air, gas in general is commonly needed in industrial processes and in buildings. A great amount of energy is consumed world-wide with conventional approaches. In this presentation, a new, versatile membrane technology is introduced to reduce energy consumption and capital cost for these applications. The membrane is prepared by deposition of a water-selective molecular sieve (zeolite) membrane on a thin (50um), robust porous metal sheet support. Such a membrane provides a unique combination of performance attributes that could not be obtained with conventional membranes, such as mechanical strength, thermal and chemical stability, high thermal conductivity, and high water permeance. The membrane can be used to separate water vapor from a hot gas stream by pervaporation and result in more than 50% energy saving for air conditioning of buildings in hot humid climate. The membrane can also be used to conduct humidity and/or heat exchange for energy recovery of building ventilation. The exceptionally high water permeance (>1.0E-5 mol/m2/s/Pa) of this membrane dramatically reduces usage of membrane surface area and enables development of a compact, cost-competitive membrane device suitable for these applications involving large volume of gas flow.

# (220c) Application of Multi-Element Structure Array On Activated Carbon Filter in HVAC Systems

Guomin Xu, Chemical Engineering, Auburn University, Auburn, AL and Bruce Tatarchuk, Department of Chemical Engineering, Auburn University, Auburn, AL

Activated carbon is widely used in industries to remove gas phase contaminants such as ozone and volatile organic compounds (VOC). Coarse particles are mostly used in commercial carbon filters mainly due to their low flow resistance despite the fact that they have a low mass transfer rate which leads low bed utilization and short breakthrough times. In contrast, fine particles have higher mass transfer rate but are limited by high pressure drop. This study aims to improve the carbon filter to obtain a high carbon loading while maintaining a low pressure drop. To reach this goal, a multi-element structure array (MESA), which combines multiple filters into single housing, was developed. A W-shape MESA bank can obtain a tenfold resistance reduction while introducing fourfold carbon load. A predictive pressure drop model for single honeycomb carbon filters, as well as MESA bank, was derived based on fundamental fluid dynamics equations and experimental data. The model can be used as a design tool to optimize carbon filter's parameter including filter depth, particle size, and filter count. Microfibrous entrapped sorbent (MFES) was prepared by entrapping fine particles (180-300μm) into 3D sinter-locked network of polymer or metal microfibers via a wet-lay process. These materials excel in mass transport and pressure drop due to their uniform structure with small particles and adjustable porosity. A polishing layer (a thin layer of MFES media placed in series with the packed bed) was used to improve the performance of bed by "polishing" the trace volume of gas after packed bed. Hexane adsorption breakthrough test was used to evaluate the performance of the composite bed based on the breakthrough time and percentage of adsorbent utilization. Experimental breakthrough curves showed that composite bed outperformed packed beds in single pass remove efficiency, breakthrough time and bed utilization due to a synergic effect between the high capacity of packed bed and the high contacting efficiency of MFES layer.

## ENERGETIC AND EXERGETIC PERFORMANCES OF A SOLAR STORAGE COLLECTOR BASED ON CERAMICS MATRIXES

**Leonardo Alonso**[1], Mario R Bernal[2], Gabriel Camargo[2] and Fabio Sierra[3], (1)Agroindustrial Engineering, Corporación Universitaria del Meta, Villavicencio , Colombia, (2)Mechanical Engineering, Universidad Libre, Bogota, Colombia, (3)Mechanical Engineering, Universidad Nacional, Bogotá, Colombia

## INTRODUCTION

In these times, the interest for materials and processes that integrate energy and economy are driven different studies in the solar collector fields. The commercial solar water heating systems (SDHW) have high investment cost which requires high initial investment and auxiliary energy used to compensate the solar energy during non sunny days. In Tunisia, Hazami et al, (1) designed and built a low cost new solar collector used for domestic water heating called integrated solar storage collector (ISSC) and characterized by an absorber matrix made up of a thin cement concrete slab which performs the function of both absorbing and storing of the solar thermal energy. There are other similar researches (2-6) looking for more economics systems of supply hot water for the population in different parts of the world. The (ISSC) technology has several advantages, if they are used in non developing countries:

a- Inexpensive, it does not require a storage tank (main cost factor).
b- Local production from available materials
c- Do not use auxiliary equipment for control purposes.

In the present study, a ISSC was built using two different ceramics materials as absorber matrices and the absorbers was tested under controlled laboratory experimental conditions. The data generated were statistically first and then in thermodynamics form.

Two solar collectors were built with 0.325 $m^2$ of area, and using as absorber matrix made up of a thin ceramics slab which performs the function of both absorbing and storing solar thermal energy. Two different commercial ceramics materials were used for each solar collector. The ceramics used are called red and white ceramic. Each solar collector has a glass, a container, polyurethane foam as insulating material and finally a metallic frame.

## EXPERIMENTAL PROTOCOL

In order to achieve the objectives of the investigation, an experimental design was established. The ISSC were tested in laboratory environment and with the following conditions;

a- Incandescent lamps were using to simulate the solar radiation. Three levels of radiation were selected 400, 500 and 750 $w/m^2$

b- The water initial temperature was always the temperature at which the aqueduct water supplied in Bogota.(16 C ).

c- The duration for each experiments were 4 hours

d- All measurements ended after 120 minutes of heating and turn off the lamps, the temperature sensors followed recording data until the initial condition was reached again

The incident radiation was measure with a DATALOGGER LI – 1400, and the acquisition a data collection systems were recorder with a NI FP 1000 equipment.

The following experimental procedure were used

1- Fill the containers (0.125 L, for each collector).
2- To put ISSC(with the thermocouples) parallel to radiation panel (0.5 m)
3- Turn on the radiation panel, with the incident radiation level chosen.
4- Recorder data.

The figures 1 and 2, shown the thermocouples position in the system and the solar collector structure.

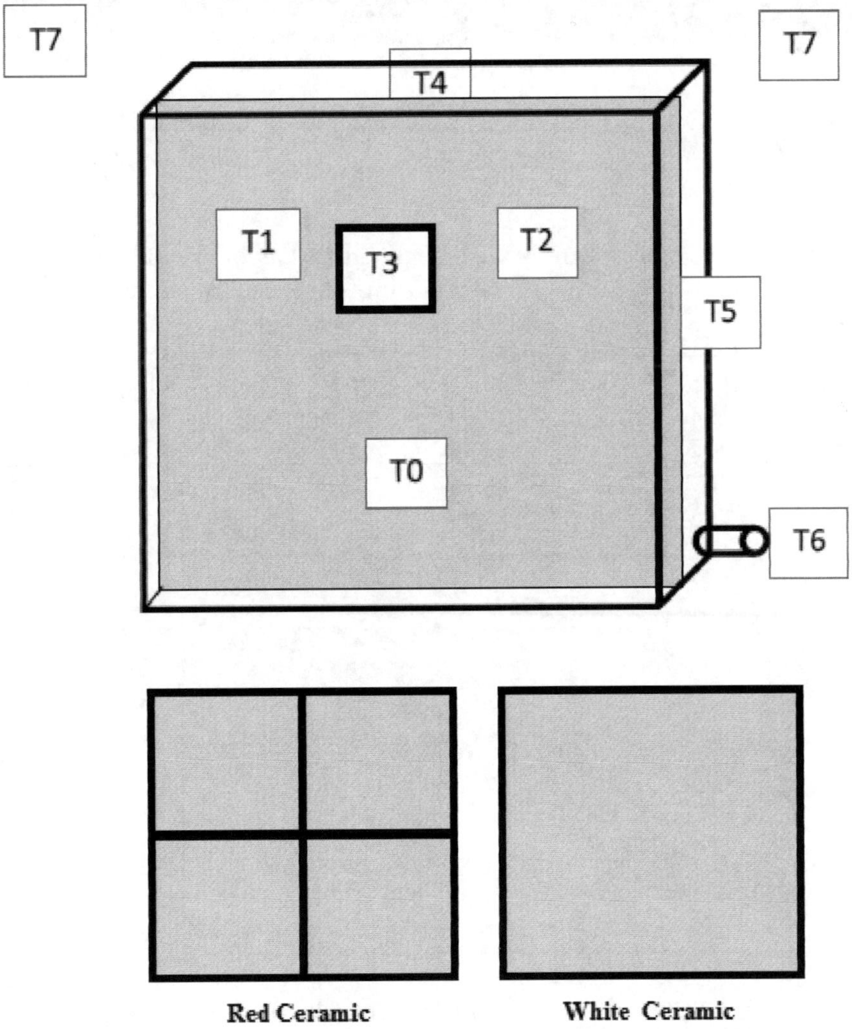


Figure 1. Thermocouple positions for the ISSC experiments

The temperature 4 was behind the collectors and the temperature 6 was for water inside the containers. The temperature 5 registered the temperature of a side of the ISSC and the thermocouple number 7 corresponded to the room temperature (environmental temperature).

The solar collectors' structure is shown in the figure 2. The ISSC in this study is a close system.

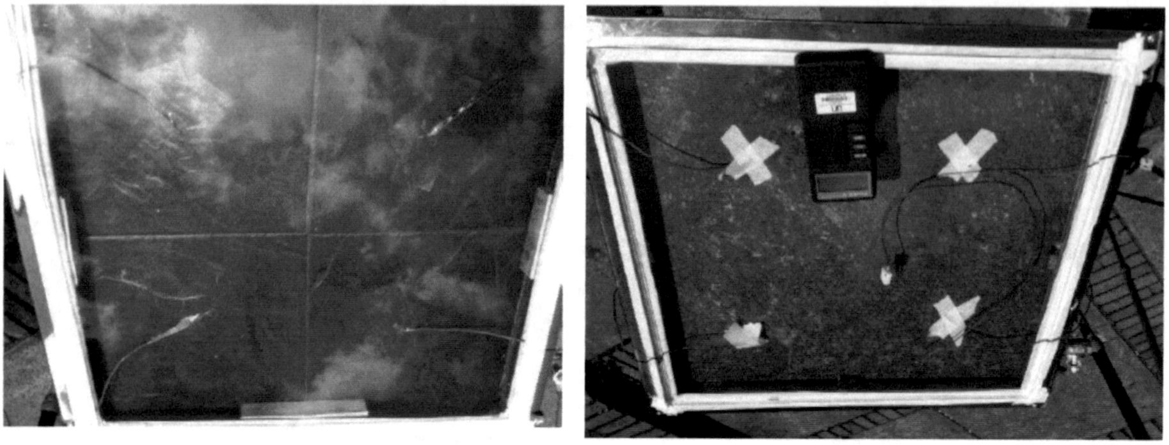

**Figure 2**. Solar Collector with ceramic materials as absorber

## ENERGY AND EXERGY ANALYSIS

The energy analysis presented in this section is mainly based on the first law of thermodynamics and the main goal of this part is to know the energy taken by the water during the experiments for calculate the energy efficiency of the system.

The theoretical model used for the study of (ISSC) by Polo Bravo (7) is an energy balance on the collector as viewed in figure No. 3

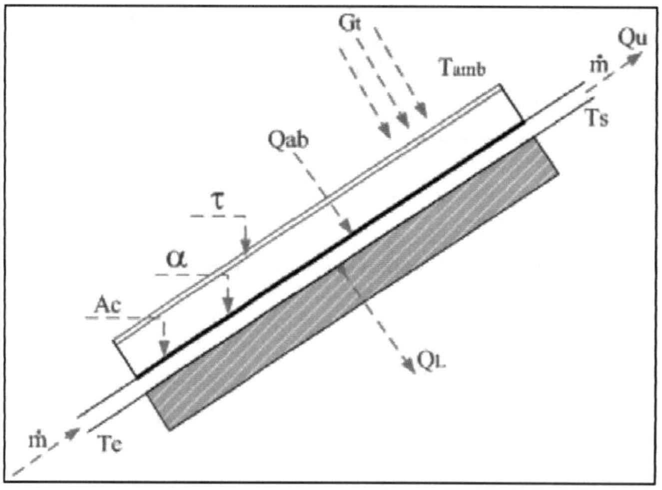


Figure No. 3: ISSC scheme with experiment energy balance (Polo Bravo)

**Nomenclature:**

$Q_U$: [W] useful power

$Q_L$: [W] power lost by radiation and convection

Qab :[W] power absorbed by the ceramic

Te,Ts :[°C] input temperature, output temperature

Tamb. [°C] environment temperature

M: [kg/s] mass flow of heat transfer fluid

Gt: [$W/m^2$] surface radiation

Ac: [$m^2$] absorber surface area

$U_L$: [W,/$m^2$ K] Global loss coefficient

$\propto\tau$: transmittance absorptance product

Energy flows mentioned are defined by:

QU = mc (Ts-Te), [1]

QL = ULAc (Tm - Tamb ), [2]

Tm = (Te + Ts)/2 [3]

Qab = $\tau \propto$Gt Ac [4]

The definition of the instantaneous efficiency $\eta$ is:

$\eta$ = QU / Gt •Ac = m•c• (Ts - Te) / Gt •Ac = Useful energy / incident energy [5]

The parameters, mass (m), specific heat (c) of the sensor must be known, while Ts, Te, AC and GT must be measured experimentally.

The useful energy (QU) can also be defined as the difference between the energy absorbed (QAB) and the losses of the collector (QL). If all the solar radiation absorbed and converted into heat is transferred to the fluid (heat transfer factor FR = 1), then:

QU = Qab – QL = $\tau$•$\propto$•Gt •Ac - UL •Ac - (Tm - Tamb) [6]

Dividing by (Gt • Ac), we obtain the equation:

$\eta = \tau \cdot \alpha - UL\ (Tm - Tamb)/\ Gt$ [7]

Is known as the characteristic equation of the instantaneous efficiency flat plate collector, and can be plotted as a graph on a straight $\eta = f\ [(Tm\text{-}Tamb)\ /\ Gt]$. In mathematical interpretation of this line, $\tau \bullet \alpha$: is the intersection with the y-$\eta$, and UL is the slope of the curve:

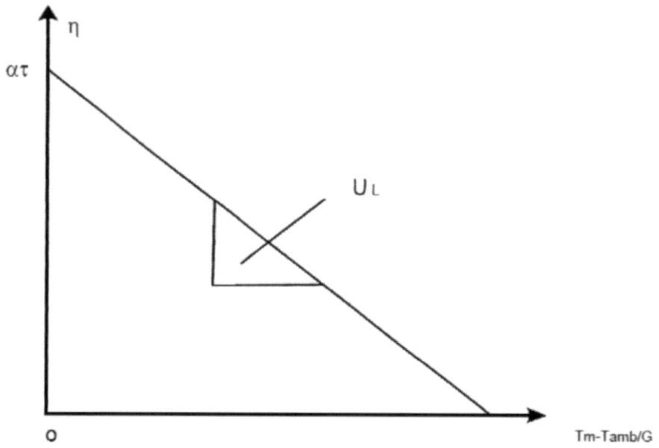


Figure 4 Characteristic of the instantaneous efficiency of a flat plate collector according to operating conditions Tm, Tamb and Gt.

## RESULTS

The different results are shown in the figures No 5 to No 10, where the temperatures for each part of the solar collector vs. time are drawn. When these data are used in the energetic analysis the global efficiency and the instantaneous efficiency are evaluated. For all levels of radiation the white ceramics materials presents a higher efficiency than the red one material. The energy is stored in the absorber materials and then transferred to water when the lamps are turn off.

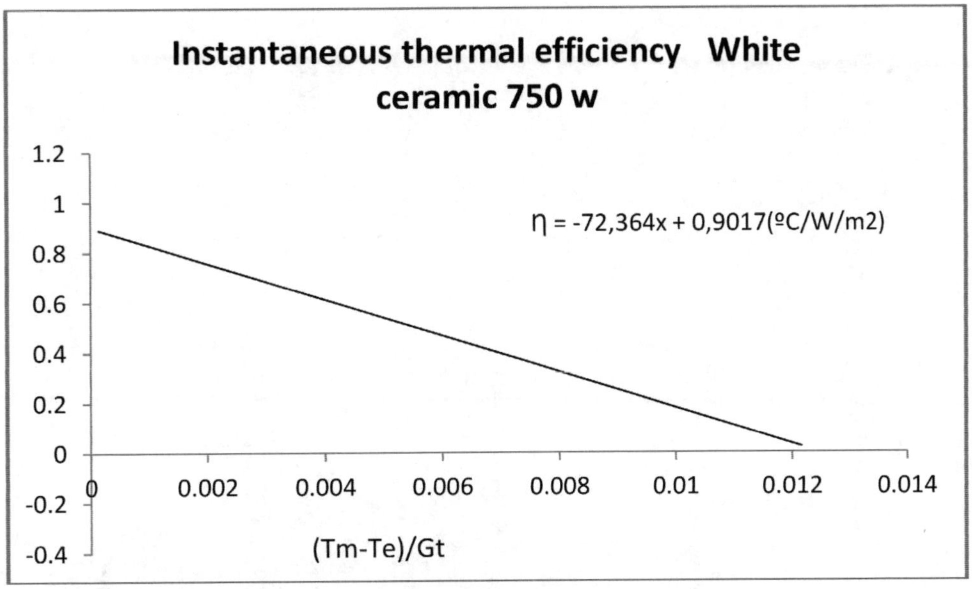


Figure 5 Instantaneous thermal efficiency curve 750W

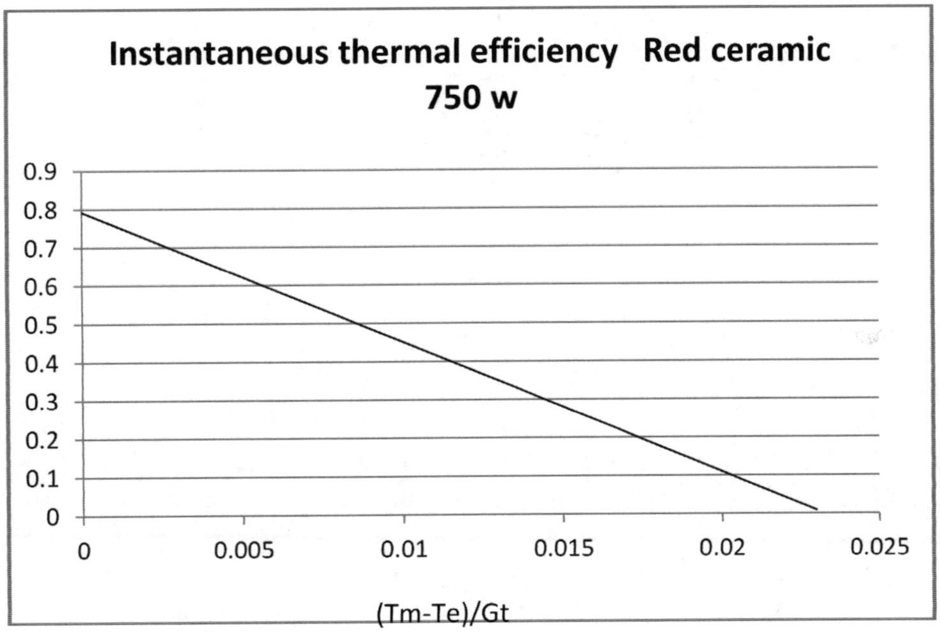


Figure 6 instantaneous thermal efficiency curve 750W

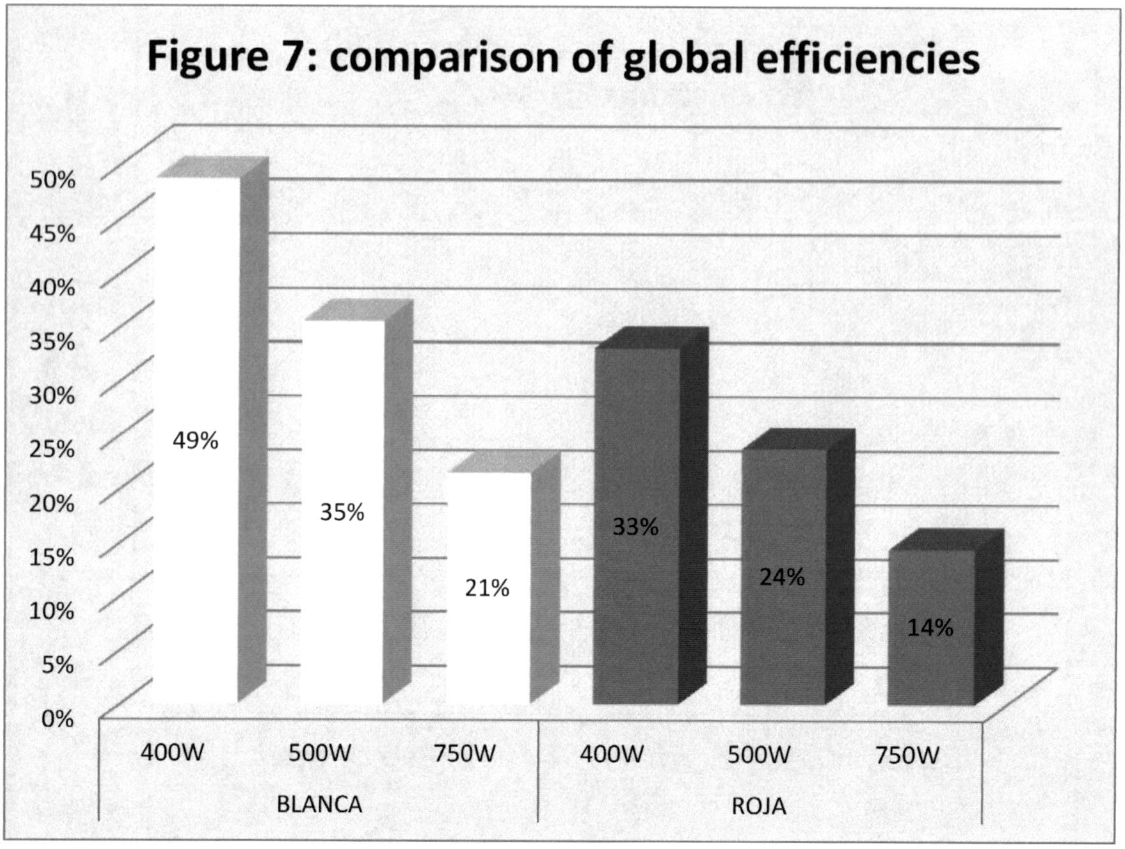


REFERENCES


1. Majdi Hazami, S. K. (29 de Enero de 2010). energetic and exergetic performances of an economical and available integrated solar storage collector based on concrete matrix. *Energy Conversion and Managment.*
2. C. Dharuman *, J. A. (2006). Performance evaluation of an integrated solarwater heater as an option for building energy conservation. *Energy and Buildings 38 (2006) 214–219*(38), 214-219.
3. Callister, W. D. (1996). *Introducción a la ciencia e ingeniería de los materiales, Volumen 2.*
4. Casanova Colás, J. (1993). *Curso de Energía Solar.* Valladolid: Universidad de valladolid.
5. Cengel, Y. A. (2006). *Thermodynamics: An Engineering Approach, 5th ed.* New York: McGraw-Hill:.
6. D. Luna a, *. J.-P. (2009). Solar timber kilns: State of the art and foreseeable developments. *Renewable and Sustainable Energy Reviews*, 1446-1455.
7. D.U. Tulyaganov a, b. S. (2009). Synthesis and properties of lithium disilicate glass-ceramics in the system $SiO_2$–$Al_2O_3$–$K_2O$–$Li_2O$. *Ceramics international*, 3013-3019.
8. Division of Agricultural Engineering and Energy Management, Central Arid Zone Research Institute, Jodhpur. (1999). Solar water heaters based on concrete collectors. *Energy*, 703-716.

9. Ebru Kavak Akpinar *, F. K. (2010). Energy and exergy analysis of a new flat-plate solar air heater having. *ELSEVIER*, Article in press.
10. Energy Information Administration. (2010). *International Energy Outlook.*
11. Exiplast S.A. (2010). Cotizacion.
12. Galimberti, P. (2009). *diseño de laboratorio de ensayos fotovoltaicos.* Andalucia.
13. Heim*, D. (2010). Isothermal storage of solar energy in building construction. *Renewable energy*, 788-796.
14. Instituto de Tecnológia cerámica. (24 de Junio de 2011). *Tecnociencia.* Recuperado el 24 de Junio de 2011, de http://www.tecnociencia.es/entidades/itc/tcn_itc5.html
15. Iturrate, E. (1998). Curso básico de teleteccion con ENVI. Buenos Aires.
16. Jorge L. Alvarado a, *. W. (2008). Passive cooling systems for cement-based roofs. *Building and Environment*, 1869-1875.
17. JUAN CARLOS BURBANO J., Á. H. (Agosto de 2006). DISEÑO Y CONSTRUCCIÓN DE UN CALENTADOR SOLAR DE AGUA OPERANDO POR TERMOSIFON. *Scientia et Technica* ( 31).
18. JUAN CARLOS BURBANO J., Á. H. (2007). ENSAYOS Y PRUEBAS DE UN CALENTADOR SOLAR DE AGUA OPERANDO POR TERMOSIFON. *Scientia et Technica*(34).
19. L. Zhu a, R. H. (2010). Comprehensive energy and economic analyses on a zero energy house versus a conventional house. *ELSEVIER*, 1043-1053.
20. Llesa, L. B. (s.f.). Materiales no metálicos resistentes a la corrosión.
21. Marc Medrano, A. G. (2010). State of the art on high-temperature thermal energy storage for power generation. *Renewable and Sustainable Energy Reviews*, 56-72.
22. Marwa M. Hassan a, Y. B. (2006). Design, construction and performance prediction of integratedsolar roof collectors using finite element analysis. *Science Direct*, 1069-1078.
23. Murcia, H. R. (2008). Desarrollo de la energía solar en Colombia y . *revista de ingenier ía. Universidad de los Andes*, 83-89.
24. Na Zhu, Z. M. (2009). Dynamic characteristics and energy performance of buildings using phasechange materials: A review. *Energy Conversion and Management*, 3169-3181.
25. (ojo). Propiedades termicas de metales, ceramicas y polimeros.
26. Ontiveros, R. I. (s.f.). Energias .
27. Otiveros, R. I. (2009). Energias Alternativas ¿luz natural? (94).
28. T.N.Anderson a, b. M. (2009). The effectofcolouronthethermalperformanceofbuildingintegrated solar collectors. *Solar EnergyMaterials&SolarCells*, 350-354.
29. TALLER ECOLOGISTA. (2009). *Manual de construcción de colectores solares termicos.* ROSARIO ARGENTINA.
30. Universidad Libre. (2007). *Documento registro calificado Maestria en ingenieria con enfasis en energias alternativas.* Bogotá.
31. UPME,IDEAM. (2005). *Atlas de radiacion solar.* Bogotá.
32. William F. Pickard, A. Q. (2009). Renewable an sustainable energy revews. *ELSEVIER*, 1934-1945.
33. Xinping Zhou a, b. F. (2010). A review of solar chimney power technology. *ELSEVIER*, 2315-2338.

# (220e) Boosting the Energy Efficiency of Ethylene Production

**Martin Schmid**[1], Bernhard Prettenhofer[1] and Thomas Wallek[2], (1)Institute of Chemical Engineering and Environmental Technology, Graz University of Technology, 8010 Graz, Austria, (2)Institute of Chemical Engineering and Environmental Technology, Graz University of Technology, Graz, Austria

Due to the large amount of different equipment and the complex set-up of heat exchanger-networks in order to increase the recovery of heat and cold, ethylene plants are among the most complex petrochemical processes. Changes with positive effects in one part of the process can lead to effects in other parts of the plant with negative impacts on the whole process that are difficult to detect and to quantify. Process simulation is a powerful tool to display these correlations. A detailed model of the crack-gas-separation part of an ethylene plant was generated with KBC-PetroSim® based on process flow sheets, equipment data sheets, laboratory tests of products and intermediates and process data obtained from the process control system during a three day test run under full load. The complete model contains 268 devices and 995 streams with 74 recycles. About 81% of the energy to run the gas separation is delivered by the crack-furnaces, the rest has to be imported from other sources within the refinery. The cryogenic processes for the separation of the light gases are counted among the major energy consumers and are composed of a 3-stage propylene refrigeration system and a 4-stage ethylene refrigeration system. Due to economic and technical issues the plant cannot always be operated at the design point. Especially partial load conditions are crucial for the compressors of the cryogenic systems. Recycle streams, which ensure a minimum flow rate to critical equipment cause a reduced energy efficiency of the whole process. The simulation of the plant shows how to adapt the control strategy in order to move energy demand from sections that are working under full capacity to these critical sections and obtain energy savings of 2% based on the imported energy. Using the model of the plant as a tool for pinch and capacity analysis it is possible to develop different scenarios for the modification of the heat exchanger network, which allows a higher interaction between the two refrigeration systems. The so obtained savings of energy are up to 7.6% based on the imported energy and 8938 t/a of $CO_2$ emissions.

# (220f) Automating Molecular Modeling of Gasification with the Kinetic Modelers Toolbox

**Scott R. Horton**[1], Yu Zhang[2], Craig A. Bennett[1], Frank Petrocelli[2] and Michael T. Klein[1], (1)Chemical and Biomolecular Engineering, University of Delaware Energy Institute, Newark, DE, (2)Tonnage Gases, Equipment and Energy - Technology, Air Products and Chemicals, Inc., Allentown, PA

**Automating Molecular Modeling of Gasification with the Kinetic Modelers Toolbox**

**Scott R. Horton*, Craig A. Bennett, Michael T. Klein**

*Department of Chemical & Biomolecular Engineering, University of Delaware*

**Francis P. Petrocelli**

*Air Products and Chemicals, Inc.*

A molecule-based kinetics model was developed for gasification, a prevalent chemical process used in the production of syngas from a variety of feedstocks including coal, biomass, and municipal solid waste. It is envisioned that this modeling approach will have applicability across a wide range of gasification reactor types and operating conditions, and represents a more-rigorous treatment for modeling gasification than most previous efforts described in the literature, in that it is not reliant on the definition of arbitrary, "lumped" pseudo-components nor on the assumption that the reaction trajectory achieves chemical equilibrium.

The generally macromolecular feeds to a gasification unit were described, structurally, as independent molecules or a collection of various interconnected cores (following either a polymeric distribution or confined to a lattice). Gasification reaction processes fall into three major categories: pyrolysis, gasification, and combustion. The feed structures were subjected to a series of reaction and logistical rules via the Interactive Network Generator (INGen) software in order to obtain a reaction network comprising all three reaction classes.

Polymeric and lattice-type species were conceptually broken into "cores" and "substituents", reacted independently, and reformed into molecules by following the Attribute Reaction Model (ARM) credo. This methodology removed the permutative explosion of species from the balance equations, and placed it on the post reaction analysis. Isomeric detail was collapsed into representative species. Each species was broken into a series of composition components (such as ring, group, and element counts).

The balance equations for the reaction network were automatically generated and solved using the Kinetic Model Editor (KME) software. The specifics of the reactor design led to the modeling of interconnected ideal reactor zones, wherein the conditions of reaction could be independently varied. It was then possible to use process, experimental, and published data to optimize the kinetic parameters.

# (316a) Effects of Carbon Sources, Oxygenation and Ethanol On the Production of Inulinase by K. marxianusYX01

**Wenjie Yuan**, School of life science and biotechnology, Dalian University of Technology, Dalian, China and Jiaoqi Gao, School of Life Science and Biotechnology, Dalian University of Technology, Dalian, China

Inulinase is one of the most important factors of the consolidated bioprocessing (CBP), which combines enzyme production, inulin saccharification, and ethanol fermentation into a single process. In this article, the inulinase production and cell growth on the *K. marxinaus* YX01 under different conditions were studied. Carbon source was shown to be less significant on the production of inulinase although the activity of inulinase was a little higher using inulin as carbon source than using glucose or fructose. The concentration of the carbon sources had a great repression effect on the activity of inulinase, when the concentration is up to 60g/L, the activity of the inulinase was only 50% of that of 20g/L. The enzyme activity was strongly influenced by aeration rate. It has been shown that the activity of inulinase and cell growth under anaerobic conditions both kept at a low level but aeration at 1.0 vvm (air volume/broth volume·minute) led to higher activity. It is interestingly that the inulinase activity per unit biomass did not show significant differences under different aeration rates. Ethanol had great repression on the cell growth. The cell stopped to grow when the addition ethanol was over 9% (v/v), but ethanol did not restrain the activity of secreted inulinase and the enzyme was stable under 15% ethanol addition.

# (316b) Effects of Solvent and Enzyme Source for Transesterification of Waste Oils

Michael Gagnon[1], Jonathan Fogel[2], Subhash Minocha[2] and **Palligarnai Vasudevan**[1], (1)Chemical Engineering, University of New Hampshire, Durham, NH, (2)Biological Sciences, University of New Hampshire, Durham, NH

Enzymatic transesterification with lipase as the catalyst eliminates soap formation. Unlike alkali-based reactions, the products can easily be collected and separated. Moreover, enzymes require much less alcohol to perform the reaction, and can be reused despite some loss in activity at the end of each cycle. Several strains of lipases have been found to have transesterification activity, namely *Candida antarctica*, *Pseudomonas cepacia*, and *Thermomyces lanuginosus*. The two primary obstacles in enzyme-catalyzed reactions are i), the immiscibility of the two substrates, namely hydrophilic methanol and hydrophobic triglyceride, resulting in the formation of an interface leading to mass transfer resistance; and ii), the strong polarity of methanol, which tends to strip the active water from the enzyme's active site leading to enzyme deactivation. The addition of an organic solvent as the medium to the reaction system simultaneously overcomes the two limitations by enhancing the solubility of oil and methanol in the solvent, and by limiting the concentration of methanol surrounding the enzyme.

We are genetically engineering plants to constitutively express a lipase for biodiesel production from spent oils. We have cloned the gene of a lipase with known transesterification activity from *Thermomyces lanuginosus*, a thermophylic fungus. Cloning involved isolation of total RNA, reverse transcription of the mRNA into cDNA and PCR amplification of the lipase gene using specific primers. The gene was first inserted into a cloning vector (pCR8/GW/TOPO) and sequenced to confirm its identity. The gene has been inserted into a plant destination vector (pGWB408 and pMDC83) via LR clonase reaction. *Nicotiana tabacum* (tobacco) leaf was transformed with the lipase gene using *Agrobacterium tumefaciens* (strain GV3101) and transferred onto selection media plates. We have also grown *Arabidopsis thaliana* from seeds that were transformed using the floral dip transformation method. The recombinant enzyme was collected from the genetically engineered plants, purified, and tested for both hydrolytic and synthetic activity. The activity results will be compared with enzyme catalysts from a commercial *Thermomyces lanuginosus,* and the effect of solvent will also be presented.

# (316c) Production of Renewable Fuel From Enhanced Activated Sludge Through a Fluidized-Bed Catalytic Cracking (FCC) Process

**Emmanuel Revellame**, William Holmes, Rafael Hernandez and W. Todd French, Dave C. Swalm School of Chemical Engineering, Mississippi State University, Mississippi State, MS

Annual production of activated sludges in the United States ranges from 7 – 8 million tons. Currently, disposal technologies of these sludges include incineration and landfilling, both of which are heavily regulated and could create adverse environmental impacts. A fraction of the activated sludges can be used as fertilizer once it complies with regulations for Class A biosolids. Activated sludges contain microbial consortia known to produce lipidic materials and many other organic compounds, which could be used as fermentation seed for enhancement of its lipid content. Several studies showed that reasonable yields (2 – 6% weight) of biodiesel could be produced from activated sludge. At these yields however, activated sludge biodiesel is less economical than petroleum diesel. By increasing the fuel yield, which could be accomplished by enhancement through fermentation, in combination with an in situ catalytic conversion to fuel, the economics of this feedstock could be improved. This study describes the production of renewable fuel from enhanced activated sludge using Fluidized Catalytic Cracking (FCC). This process could convert most of the compounds present in sludge into fuel, and thus, increase fuel yields and reduce cost.

# (316d) Transformative Catalytic Technology for Advanced Biofuel Production

Manhoe Kim, NextCat Inc, Detroit, MI, Like Yang, NextCat INC, Detroit, MI, Steven O. Salley, Chemical Engineering and Materials Science, Wayne State University, Detroit, MI and K. Y. Simon Ng, Department of Chemical Engineering and Materials Science, Wayne State University, Detroit, MI

Biodiesel producers are facing challenges of high cost feedstocks and lack of enabling technology to process low grade (low cost) lipids having high free fatty acid (FFA) and water levels. In order for biodiesel producers to became commercially viable, they must be able to use less expensive lipid feedstocks, a wide range of crude bio-oil feedstocks, with varying FFA content, including palm oil (FFA ~12%), residual corn oil (FFA ~15%) from the distiller dried grains (DDGs), animal fats (FFA 2~35%), yellow grease (FFA 5 - 15%), and brown grease (50~100%). In order to avoid the problems related to homogeneous catalytic system, a transformative heterogeneous catalytic conversion technology using the heterogeneous metaloxides catalyst supported by $ZrO_2$ has been studied in continuous flow reactor systems. The catalyst processing factors (aging, drying, and extruding, and drying) and the reactor scale-up factors (reactor diameter and length ratio, residence time, superficial flow velocity) which determine catalytic activity and durability has been studied. Downward feed flow has a plug flow feature and gives higher FAME yield than upward flow mode. Superficial velocity of the feed gives effects on both the degree of mixing of the reactants and dissolution of the active sites on the catalyst surface due to shear force induced by the viscous flow. In order to decrease shear force, the void space of the catalyst bed should be increased and the reactor diameter and length ratio should be optimized. This technology can potentially reduce the feedstock, operating and processing costs and reenergize the latent biodiesel production facilities to enable to be used for biodiesel production with various non-edible waste lipids.

# (316e) Transesterification of Triglycerides for Biodiesel Over Selected Zeolite Catalysts

Yu-Yuan Wang, Hsin-Yu Chou and **Bing-Hung Chen**, Department of Chemical Engineering, National Cheng Kung University, Tainan, Taiwan

Transesterification of triglycerides in excess methanol to biodiesel using selected zeolite catalysts has been carried out. Various zeolites, such as beta-zeolite, zeolite Y and Mordenite, doped with alkali metals either by ion-exchange or impregnation, are employed to catalyze the transesterification at a relatively lower temperature around 60°C. A faster transesterification rate and higher conversion efficiency would be achieved with appropriate pretreatment on catalysts. Under a reaction temperature at 60°C, conversion efficiency of triglycerides to biodiesel as high as 97.3% was obtained. Cyclic durability of zeolite catalyst has been examined by running the same zeolite catalysts for 6 cycles of transesterification of triglyceride in methanol. It is found that catalyzed conversion efficiency of triglycerides did not differ significantly during the first 3 cycles. Additionally, regeneration on spent zeolite catalysts has been attempted as well. The zeolite catalyst can be recycled at least for four times without significant deterioration in terms of conversion efficiency of triglycerides.

# (316f) Hydrotreating of Soybean Oil Over NiMo Carbide Catalyst On Five Different Supports

Huali Wang[1], Shuli Yan[2], **Elvan Sari**[1], Steven O. Salley[1] and Simon K. Y. Ng[1], (1)Chemical Engineering and Materials Science, Wayne State University, Detroit, MI, (2)NextCAT inc., Detroit, MI

**Abstract**

As an alternative to conventional sulfide catalysts, NiMo carbide catalysts were prepared by precipitation method and studied for the hydrotreating of soybean oil to produce hydrocarbons in the gasoline to diesel range. The effect of the catalyst supports on activity and selectivity was investigated by using three different types of materials: mesoporous material (Al-SBA-15), alumina ($\gamma$-$Al_2O_3$) and zeolites (ZSM-5, Zeolite $\beta$ and USY). The structural characterization of the catalysts was performed using XRD, BET and TEM. Catalytic tests were carried out in a bench scale flow reactor at 400 °C and 650 psi. The results showed that the Al-SBA-15 supported catalyst has the highest surface area of 711.5 $m^2\ g^{-1}$ and largest pore volume of 0.71 $cm^{-3}\ g^{-1}$. Among the five catalysts, hydrotreating on NiMoC/Al-SBA-15 gave the highest yield of organic liquid product (96%) and highest selectivity (97%) to hydrocarbons in the boiling range of the diesel fraction. For the three zeolite supported catalysts, hydrotreating of soybean oil produced more hydrocarbon products in the boiling range of green gasoline (about 15-40%).

# (321a) Microchannel Rapid Temperature Swing Adsorption of CO2 for Methane Production

**Kriston Brooks**[1], Scot Rassat[2] and Paul Humble[2], (1)Energy and Environmental Directorate, Pacific Northwest National Laboratory, Richland, WA, (2)Pacific Northwest National Laboratory, Richland, WA

A rapid thermal swing adsorption system combined with Sabatier reactor was developed to demonstrate the applicability of a NASA microchannel Mars *In Situ* Propellant Production (ISPP) system. A thermal swing adsorption pump was designed to collect and compress $CO_2$ by rapidly temperature cycling eight cells filled with Molesieve 13X. Recuperation of energy between cells was used to minimize heat and cooling requirements. The collected $CO_2$ can be combined hydrogen from electrolyzed water to produce methane fuel using a microchannel Sabatier reactor.

Tests were performed to compress $CO_2$ by 10 fold using a feed stream of 10% $CO_2$ and 90% $N_2$ at 1, 2, and 4 minute cycle times. With the initial testing, a compression ratio of 5.3 was achieved with 71 to 79% thermal recuperation efficiency. Subsequent modeling work has refined the design and the number of cells required to improve the overall system efficiency and simplify the process. These results indicate similar thermal efficiencies can be achieved using a system with only four cells rather than eight. Modeling results further demonstrate design changes required to obtain improved efficiencies and compression ratios. Results of both the original experimental work and the more recent modeling efforts will be presented.

# (321b) Solubilities of CO2 in Aqueous Solutions of Ionic Liquid and Monoethanolamine

**Fan Xu**[1], Haifeng Dong[2], Xiangping Zhang[2], Hongshuai Gao[3], Zhanli Wang[1], Suojiang Zhang[2] and Baozeng Ren[4], (1)Institute of Process Engineering, Beijing, China, (2)Institute of Process Engineering, Chinese Academy of Sciences, Beijing, China, (3)State Key Laboratory of Biochemical Engineering, Key Laboratory of Green Process and Engineering, Institute of Process Engineering, Chinese Academy of Sciences, Beijing, China, (4)Zhengzhou University, Zhengzhou, China

With the deterioration of climate change, $CO_2$ capture and sequestration have been paid much attention to reduce greenhouse gas emissions. Currently aqueous solutions of alkanolamines or their mixtures are commonly used in many industrial processes to capture $CO_2$. However, drawbacks such as solvent loss and high regeneration costs due to the high water content drive researchers for new superior technologies [1]. Ionic liquids (ILs), due to their wide liquid range, low point, tunable properties, negligible vapor pressure, high $CO_2$ solubility and reasonable thermal stability, are emerging as promising candidates to capture $CO_2$, but it is difficult to realize industrialization owing to its high viscous and high cost. Recently, the idea of mixing ILs and alkanolamines has been received great attention from the industries, since its advantages and capabilities in reducing the problem caused by the usage of conventional alkanolamines solutions are promising [2-4]. In this work, a low viscous IL ($[C_2OHmim][N(CN)_2]$) has been synthesized and the solubilities of $CO_2$ in aqueous mixture of IL-MEA were measured by instruments with high accuracy at temperature 313.15K and 333.15K, over $CO_2$ partial pressure ranging from 100 to 1000kPa and IL concentration varying from 5% to 30%. Then the corrected PVT equation has been used to calculate the absorption of $CO_2$. The results indicate that 0.583 mol $CO_2$/mol MEA could be reached under the condition of 40°C and 100.66 kPa when appropriate amount of IL has been added, which only dropped by 1.7% when compared with 30% MEA aqueous at the same condition. The results sufficiently demonstrated that the addition of the IL can reduce regeneration energy consumption to 27% with little affecting their absorption performance.

**References**

[1] Chinn, D.; Vu, D. Q.; Driver, M. S.; Boudreau, L. C. $CO_2$ removal from gas using ionic liquid absorbents. *US Patent*, 7,527,775 B2, 2009

[2] Camper, D.; Bara, J. E.; Gin, D. L.; Noble, R. D. Room-temperature ionic liquid – amine solutions: Tunable solvents for efficient and reversible capture of $CO_2$. *Ind. Eng. Chem. Res.*, 2008, 47, 8496-8498.

[3] Zhao, Y.; Zhang, X.; Zeng, S.; Zhou, Q.; Dong, H.; Tian, X.; Zhang, S., Density,

Viscosity, and Performances of Carbon Dioxide Capture in 16 Absorbents of Amine+ Ionic Liquid+ $H_2O$, Ionic Liquid+ $H_2O$, and Amine+ $H_2O$ Systems. *J. Chem. Eng., Data*, 2010, 55, 3513-3519.

[4] Ahmady, A.; Hashim, M. A.; Aroua, M. K., Absorption of carbon dioxide in the aqueous mixtures of methyldiethanolamine with three types of imidazolium-based ionic liquids. *Fluid Phase Equilibria*, 2011, 309, 76-82.

# (321c) A Nove Amino-Functionalized Ionic Liquid for CO2 Capture

**Jinzhi Zhang**, Haifeng Dong and Xiangping Zhang, Institute of Process Engineering, Chinese Academy of Sciences, Beijing, China

Ionic liquids (ILs), especially amino-functionalized ionic liquids, have been found to be effective for $CO_2$ capture because of their unique properties such as negligible vapor pressure, high thermal stability and good absorption capacity[1-2]. As reported [3-5], ILs of more amine groups could afford higher $CO_2$ solubility. To explore deeply into this viewpoint, a new ionic liquid is synthesized and used in lab scale of $CO_2$capture experiments to identify the relationship between the structure characteristics and capture performance in terms of the absorption rate and capacity.

In present work, a novel amino-functionalized ionic liquid was synthesized and characterized by $^1H$ NMR, $^{13}C$ NMR, IR and MS. The physicochemical properties including glass transition temperature and thermal decomposition temperature were tested. Experiment result shows that this ionic liquid is capable of rapid capture 0.174g $CO_2$ / g IL at 303.15K and atmospheric pressure, which is the theoretical maximum for absorption of $CO_2$ chemically. The interactions of the ionic liquid with $CO_2$ were further investigated by comparison of both the FTIR and $^{13}C$ NMR spectra of the gas-untreated and gas-treated materials to support the experimental and the theoretical results.

Moreover, the synthesized ionic liquid could be recycled under vacuum at 100°C.

**References**

[1] J. L. Anthony, E. J. Maginn, J. F. Brennecke, *J. Phys. Chem. B* **2002**, *106*, 7315-7320.

[2] L. A. Blanchard, D. Hancu, E. J. Beckman, J. F. Brennecke, *Nature* **1999**, *399*, 28-29.

[3] Z. Xue, Z. Zhang, J. Han, Y. Chen, T. Mu, *International Journal of Greenhouse Gas Control* **2011**, *5*, 628-633.

[4] J. M. Zhang, S. J. Zhang, K. Dong, Y. Q. Zhang, Y. Q. Shen, X. M. Lv, *Chem.-Eur. J.* **2006**, *12*, 4021-4026.

[5] Y. Zhang, S. Zhang, X. Lu, Q. Zhou, W. Fan, X. Zhang, *Chemistry - A European Journal* **2009**, *15*, 3003-3011.

# (321d) Assessing Anhydrous Tertiary Alkanolamines for Gas Purifications

**David J. Heldebrant**[1], Paul M. Mathias[2], Kash Afshar[2], Satish Reddy[2], Arnold R. Smith[3], Rubin J. McDougal[4], Louis V. Jasperson[4], Phillip K. Koech[5], Mark Bearden[6] and Charles J. Freeman[1], (1)PNNL, Richland, WA, (2)Fluor Corporation, Aliso Viejo, CA, (3)Fluor Corporation, Aliso Viejo, WA, (4)Wiltec Research Company, Provo, UT, (5)Pacific Northwest National Laboratory, Richland, WA, (6)Energy & Environmental, Pacific Northwest National Lab, Richalnd, WA

**Assessing Anhydrous Tertiary Alkanolamines for Gas Purifications**

Paul M. Mathias, Kash Afshar, Satish Reddy, Arnold R. Smith, Rubin J. McDougal
Louis V. Jasperson, Phillip K. Koech, Mark D. Bearden, Charles J. Freeman, David J. Heldebrant

Anhydrous tertiary alkanolamines chemically react with $CO_2$ and $H_2S$, with greater selectivity for the latter. This is in direct contrast to aqueous amine-based solvent systems with higher selectivity for $CO_2$ due to stronger chemical fixation than for $H_2S$. Anhydrous tertiary alkanolamines exhibit pressure-induced chemical fixation of $CO_2$ to form zwitterionic ammonium alkylcarbonate ionic liquids, while the same tertiary alkanolamines react with $H_2S$ at atmospheric pressures to form hydrosulfide ionic liquids. This difference in capture pressure implies that anhydrous alkanolamines could be chemically selective for $H_2S$ over $CO_2$ under specific conditions. We present here the first vapor-liquid equilibrium (VLE) data of anhydrous ethyldiethanolamine (EDEA) with $CH_4$, $C_3H_8$, $H_2S$ and $CO_2$ at 10 and 50 °C measured by the PTx and PTxy methods. The data are modeled in Aspen Plus® using an electrolyte NRTL model. Data trends and phenomena are discussed for each gas. We also compare anhydrous EDEA's simulated performance for $CO_2$ and $H_2S$ separations to other solvents such as Fluor Solvent, Econosolv, and aqueous methyldiethanolamine (MDEA) for representative gas-processing applications.

# (321e) Development of a Novel Carbonate Absorption Process with Crystallization-Enabled High Pressure Stripping for Post-Combustion CO2 Capture: Kinetic Study of Bicarbonate Salt Crystallization

**Qing Ye**, University of Illinois at Urbana-Champaign, Champaign, IL

Development of a Novel Carbonate Absorption Process with Crystallization-Enabled High Pressure Stripping for Post-Combustion $CO_2$ Capture: Kinetic Study of Bicarbonate Salt Crystallization

Qing Ye[1,2], Manoranjan Sahu[2], Yongqi Lu[2] and Xinlei Wang[1]

1 Agricultural and Biological Engineering, University of Illinois at Urbana-Champaign

2 Illinois State Geological Survey, University of Illinois at Urbana-Champaign

A novel absorption process with high pressure stripping is being developed at our group for the post-combustion $CO_2$ capture application. The process employs a carbonate salt aqueous solution, such as potassium carbonate (PC), as a solvent. The $CO_2$ absorption operates at elevated temperature and the bicarbonate in the $CO_2$ rich solution from the absorber is crystalized in a separate cooling-crystallizer. Slurry of bicarbonate crystals is used in the stripper for $CO_2$ stripping at high pressure. Compared to conventional mono-ethanol-amine (MEA) processes, the carbonate-based process consumes less stripping heat and $CO_2$ compression work. A preliminary techno-economic analysis reveals that the energy use of the novel process is 40-50% less than that of its MEA counterpart.

Crystallization of the bicarbonate from the $CO_2$-rich PC solution is a key step in the carbonate-based process. Kinetic studies of the bicarbonate crystallization were performed under continuous mix suspension, mixed product removal (MSMPR) mode using an automated calorimetric reactor system. Parametric tests were conducted to investigate the effect of the PC concentration, $CO_2$ loading, addition of absorption promoter, temperature, residence time, stirring speed, seeding, etc. on the kinetics of the crystallization and morphology of crystal products. Crystal solids were characterized by particle size distribution (PSD), turbidity, SEM, and XRD analysis techniques. Kinetic data, such as the nucleation rate constant, size-dependent growth rate constant, and agglomeration/breakage factor, were obtained by fitting the differential PSD data under various operating conditions.

This presentation will provide a summary of the results from the crystallization kinetic study and discuss their implications for the crystallizer design.

**Doped lithium orthosilicate for absorption of $CO_2$ at steam atmosphere with low $CO_2$ partial pressure**

**Abstract:** Calcium and lithium containing oxides are regarded as options for the capture of $CO_2$ in sorption-enhanced steam reforming of methane (SEMSR) system. Among the latter, lithium orthosilicate (Li4SiO4) is considered as one of the most potential materials due to its large capacity, high selectivity and good cycle performance. Improved properties can be obtained by appropriately doping different elements into $Li_4SiO_4$. In this study, the best amount of doping elements (K, Fe) into $Li_4SiO_4$ were calculated by Factsage 5.5. Therefore, different lithium orthosilicate based absorbents, potassium doped and iron doped lithium orthosilicate according to the calculated results from Factsage 5.5 were prepared and characterized by XRD, SEM. For the study of practical applicability in SESMR, the absorption ability was investigated by TGA under low $CO_2$ partial pressure (5%~20%) influenced by different steam atmosphere (5%~30%). In addition, the cyclic absorption/desorption performance of different absorbents were also investigated. The results show that potassium and iron can enhance the absorption capacity at low $CO_2$ partial pressure and cyclic ability respectively and the best absorbent with suitable amount of potassium and iron was acquired.

**Key words: $CO_2$ acceptors; Lithium orthosilicate; doping method**

## 1 Introduction

$CO_2$ acceptors are widely used in industrial process such as sorption enhanced methane steam reforming, flue gas decarburization, energy storage system and etc. Thus, it is important to obtain a solid $CO_2$ acceptor which has high $CO_2$ absorption capacity at moderate temperature with good selectivity for $CO_2$ and good regeneration ability.

Thus far, several materials, such as hydrotalcite-like materials, CaO-based sorbents, and lithium-containing materials[1], have been proposed for $CO_2$ capture. Among these materials, lithium orthosilicate ($Li_4SiO_4$) is considered to be one of the most potential materials. $Li_4SiO_4$ has been reported to adsorb $CO_2$ more than 30 times faster than $Li_2ZrO_3$ and is lighter and cheaper than $Li_2ZrO_3$.

$$Li_4SiO_4 + CO_2(g) = Li_2CO_3 + Li_2SiO_3$$

$Li_4SiO_4$ has high $CO_2$ absorption capacity. However, the temperature in the real SESRM process is between 500□~600□. Moreover, the concentration of $CO_2$ is as low as about 10%. Under such condition, the $CO_2$ absorption ability of $Li_4SiO_4$ will decrease dramatically. In order to improve the defect, exotic elements can be doped into $Li_4SiO_4$. C. Gauer [2]etc. added elements such as Al, Fe and K into $Li_4SiO_4$, concluding that for the practical use of $Li_4SiO_4$, doping vacancies through doping Fe can not only increase the sorption ability but also the cycle performance. M. Seggiani etc[3]. doped K element using simple addition method and eutectic method into Li4SiO4 and compared the two methods drawing the conclusion that simple addition method can well promote the sorption ability while has a bad cycle performance. Mayra Y. Veliz-Enriquez etc[4]. prepared $Li_{2-x}K_xZrO_3$ with different quantities of $K_2CO_3$ and then concluded that the best x is 0.2 using solid solution theory and experiment.

In the real SEMSR process, not only can the temperature reach the best absorption temperature(700□) and can the concentration of $CO_2$ be high enough for sorption, but also the steam exist to influence the sorption ability of $Li_4SiO_4$. Hughes, Manovic, Reijers, Sun, Yan, Zeman etc[5-10]. investigated the effect of steam on absorption of CaO for $CO_2$ holding the opinion that steam can facilitate the sorption ability of CaO for $CO_2$. Esther Ochoa-Fernández[11] studied the effect of steam on pure $Li_4SiO_4$ for $CO_2$ absorption but the reason was not indicated.

In this article, $Li_4SiO_4$ with different quantities of K were prepared through solid synthesis using eutectic method. The best amount of $K_2CO_3$ was calculated by Factsage software and the results were proved by characterization method such as XRD, SEM and TG. The sorption ability of different samples for $CO_2$ was investigated under the same condition as in the real SEMSR process. Finally, the effect of sorption condition on sorption ability such as temperature, pressure of $CO_2$ and the amount of steam were analyzed.

## 2 Experimental part

### 2.1 Characterization of samples

Solid-state synthesis of analytical grade lithium carbonate and highly dispersed silicon dioxide ($SiO_2$) was performed by adjusting a molar ratio of 2.05:1 to obtain $Li_4SiO_4$ powder. After homogenization in a vibrating disc mill for a minute the initial mixture was calcined in air at 750□ for 6h.

For simple addition doping, the molar ratio of starting materials was adjusted to $Li_2CO_3$:$SiO_2$:$K_2CO_3$=(2.05+A):1:(2A).

### 2.2 Characterization of samples

The sample was identified by X-ray diffraction (XRD). A diffractometer (RIGAKU D/MAX 2550 VB/PC, Japan) coupled to a copper-anode X-ray tube was used. The Kα wavelength (1.54056nm) was selected with a diffracted beam monochromator. Li4SiO4 was identified by the corresponding Joint Committee on Powder Diffraction Standards (JCPDS) in virtue of MDI Jade 5.0 software. The surface structure and particle size of the sample was characterized by scanning electron microscopy (SEM) (JEOL JSM-6360LV, Japan).

A WRT-3P scale was used to analyze the temperature ranges and kinetic behaviors of the CO2 sorption process.

## 3 results and discussions

### 3.1 Theoretic effects of doped K element on $Li_4SiO_4$

Doping substance which is able to form eutectic compound with $Li_2CO_3$ can improve the $CO_2$ sorption ability of $Li_4SiO_4$. There are two common alkali metal salt including $K_2CO_3$ and $Na_2CO_3$ that can form eutectic compound with $Li_2$CO3. Fig.1 and Fig.2 show the binary phase diagram of $Li_2CO_3$/$K_2CO_3$ and $Li_2CO_3$/$Na_2CO_3$. When the amount of $K_2CO_3$ is 0.4, the melting point of the eutectic compound is 500□. While the amount of $Na_2CO_3$ is up to 0.5, the melting point is 500□. $K_2CO_3$ and $Na_2CO_3$ can both decrease the melting point dramatically but $K_2CO_3$ is better for its smaller amount needed.

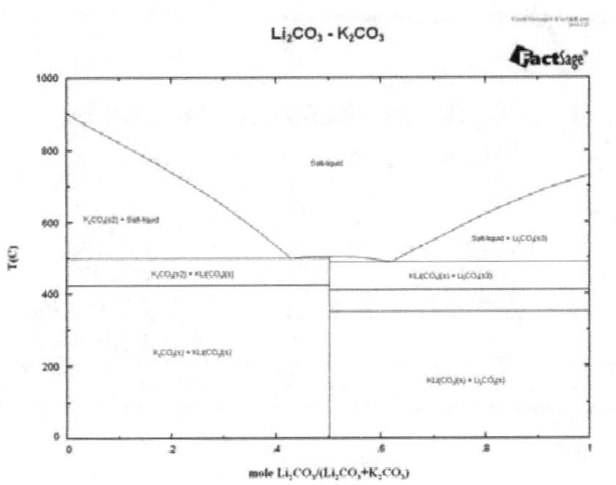


Fig.1 $Li_2CO_3$-$K_2CO_3$ binary phase diagram

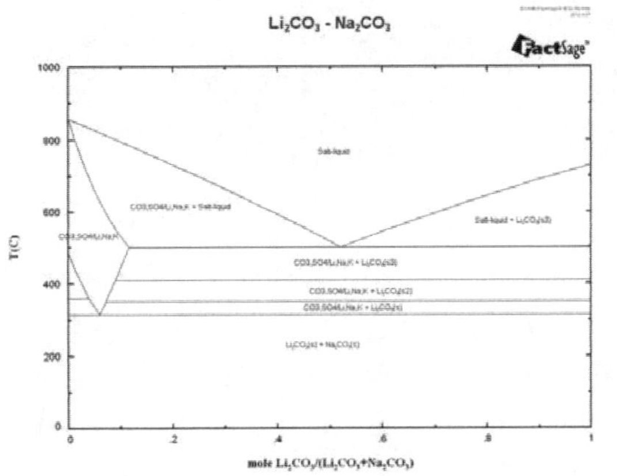


Fig.2 $Na_2CO_3$-$K_2CO_3$ binary phase diagram

Fig.3 shows the calculated equilibrium composition using Factsage software of starting materials with different amount of $K_2CO_3$

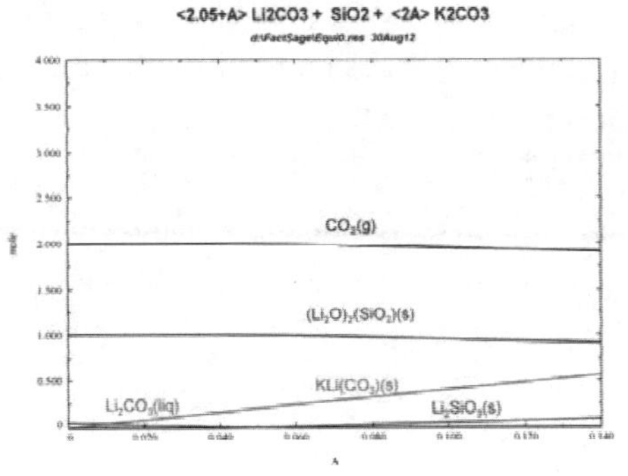


Fig.3 Calculated equilibrium composition of $Li_2CO_3$+$SiO_2$+$K_2CO_3$ calcined at 750□ for 6h

The value of A in the experiment of C. GAUER is 0.1. With the increase of A, which means the increase of amount of $K_2CO_3$, $Li_2CO_3$ that remains in the product will disappear gradually. While A increases continually, $Li_2SiO_3$ appears as a by-product, which has no promotion for the sorption of $CO_2$. A new substance, KLi($CO_3$), which is the eutectic compound formed by $K_2CO_3$ and $Li_2CO_3$, can be found with the increase of $K_2CO_3$. The melting point decrease more as the amount of $K_2CO_3$ increase thus the sorption ability for $CO_2$ is improved more. In summary, the value of A should be maintained between 0.04~0.06 so the absorbent can have more KLi($CO_3$)

while less impurity.

### 3.2 influence of amount of $K_2CO_3$ to sorption ability of K-$Li_4SiO_4$

Table.1 shows the value of A in different samples. The molar ratio of starting materials is as follows: $Li_2CO_3$:$SiO_2$:$K_2CO_3$=(2.05+A):1:(2A)

Table.1 value of A in different samples

| sample | 1 | 2 | 3 | 4 | 5 | 6 |
|---|---|---|---|---|---|---|
| A | 0 | 0.06 | 0.08 | 0.10 | 0.12 | 0.14 |

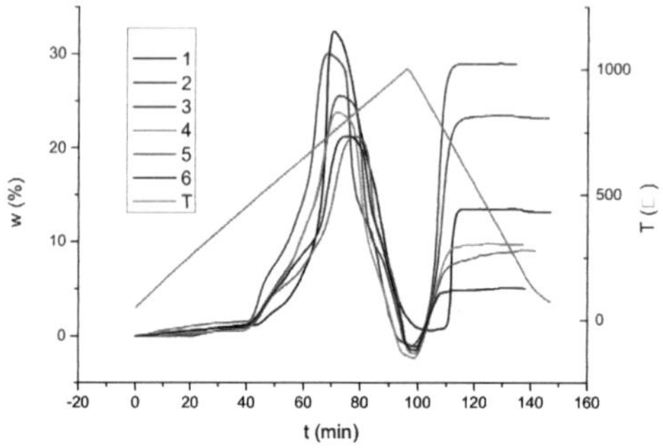


Fig.4 thermo analytical curve of different samples

The thermo analytical curve in Figure 4 was obtained by heating the samples in the pure $CO_2$ atmosphere with a heating rate of 10□/min from 0°C to 1000℃ and then with a cooling rate of 20℃/min from 1000℃ to room temperature. The samples which have $K_2CO_3$(No.2~No.6) has a clearly improvement of $CO_2$ sorption ability at 50min when the temperature is about 500□ compared with pure $Li_4SiO_4$(No.1), which indicates that the eutectic compound can indeed improve the $CO_2$ sorption ability. No.2 sample has the most significant increase at 500□. Nevertheless, the sorption ability decreases as the amount of $K_2CO_3$ increases. This can be attributed to the increase of the $Li_2SiO_3$, which has no positive effect for $CO_2$ sorption. The results are well consistent with the results calculated by Factsage before.

During the temperature decrease process, all samples absorb $CO_2$ for the second time. This process is rather short for the cooling rate is as twice as much as the heating rate. Even though, No.2 sample can have a weight gain of 30%, which is obviously better than other samples, within such a short time.

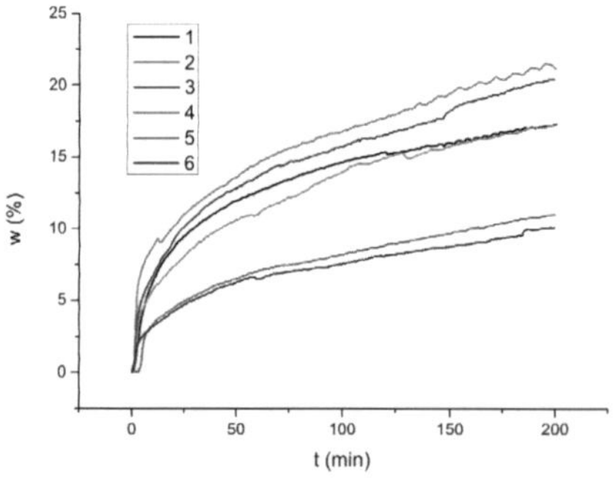


Fig.5 Isothermal curves of different samples at 575□ and 100% $CO_2$ atmosphere

In the real SESRM process, the low the temperature is, the less energy the reaction consumes. Fig. 5 shows the isothermal analysis curves of different samples at 575℃ and 100% $CO_2$ atmosphere. No.2 and No.3 samples have better sorption performance than No.1 sample and No.2 has a bigger increase. Compare No.2 sample with No.1 sample, the rate of rapid reaction stage increases. This can be attributed to the formation of eutectic compound, which restrains the formation of the double shell because liquid phase eutectic compound can facilitate the diffusion of $Li^+$ and $O^{2-}$ and thus higher conversion is needed to form the solid double shell. With the increase of $K_2CO_3$, the sorption ability, on the contrary, decreases. The reason is similar with the reason for the results of the thermal analytical curves.

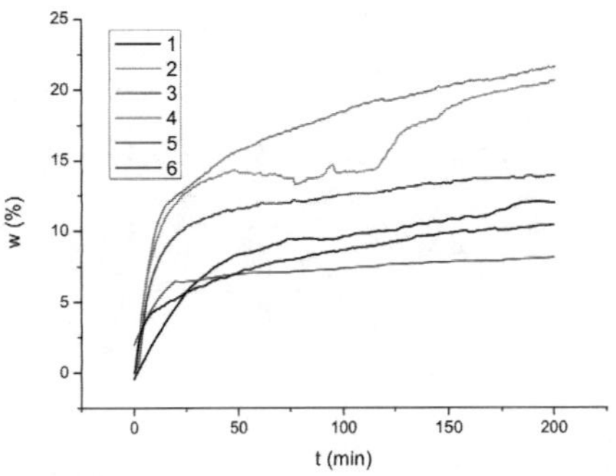


Fig.6 Isothermal curves of different samples at 575℃ and 10% $CO_2$ atmosphere

Fig.6 shows the isothermal curves of different samples at 575℃ and 10% $CO_2$ atmosphere. The rate of the rapid reaction stage of No.1 sample is significantly low than any other samples because of the low $CO_2$ concentration. While for the doped with K samples, No.2~No.6 samples, the rate of the rapid reaction stage increases apparently. This can be attributed to the formation of eutectic compound which can increase the amount of $Li^+$ and $O^{2-}$ on the surface and thus facilitating the following reaction.

$$2Li^+ + O^{2-} + CO_2 \underset{k_4}{\overset{k_3}{\rightleftharpoons}} Li_2CO_3 \qquad (3)$$

For No.2 sample, the weight gain can easily reach to 13% within 20min and to 22% within 200min, which is clearly higher than that of No.1 sample. As the amount of K increases, the sorption ability decreases. The reason is also the same as the reason for the results of thermal analytical curves.

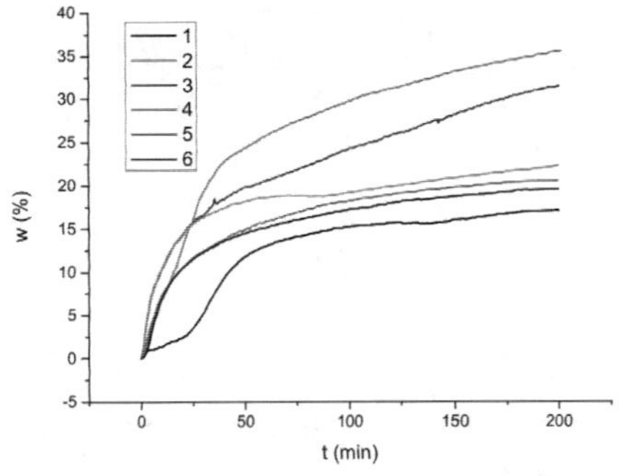


Fig.7 Isothermal curves of different samples at 575℃, 10% $CO_2$ and 20% steam atmosphere

Fig.7 shows the isothermal curves of different samples at 575℃, 10% $CO_2$ and 20% steam atmosphere. For No.1 sample, the weight gain within 200min increases from 13% to 15%, while fro No.2 sample, the weight gain increases from 22% to 35%. For the other samples, the sorption capacity also increases more than No.1 sample.

### 3.3 Characterization of samples

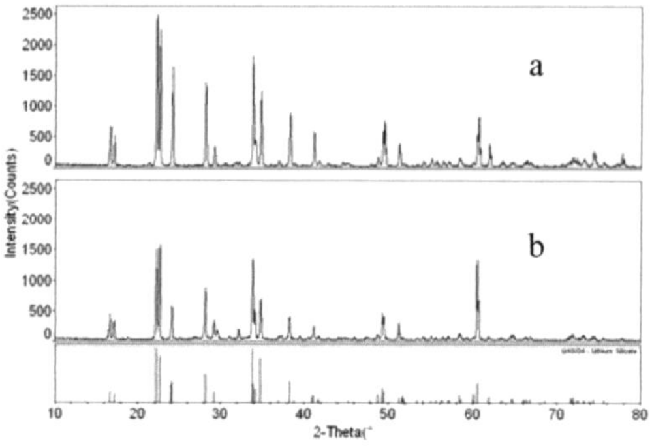

a-pure $Li_4SiO_4$; b-No.2 sample

Fig.8 XRD patterns of different samples

Fig.8 shows the XRD patterns of pure Li4SiO4 and No.2 sample. They were all fitted to JCPDS file 37-1472, and only the highly crystalline structure of $Li_4SiO_4$ was presented. This can be explained by the solid solution theory. Small quantities of $K_2CO_3$ can be dissolved into the $Li_4SiO_4$ solvent as a solid solute, so no substances including K can be found. This also explains why the sorption ability decreases while the amount of K increases.

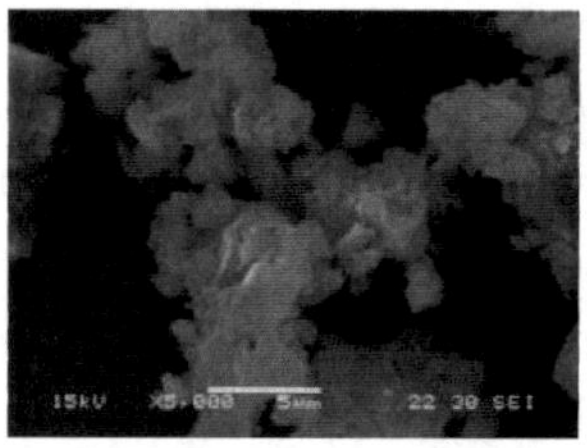

a-pure $Li_4SiO_4$

b-No.2 sample

Fig.9 SEM images of different samples

Fig.9 shows the SEM images of different samples. Pure $Li_4SiO_4$ and No.2 sample both have smooth surface and are fin powders. Pure $Li_4SiO_4$ has a smaller diameter of 10μm, while No.2 sample has an agglomeration phenomenon with a diameter of 30μm and has small particle which is $Li_4SiO_4$ with K on its surface. Thus, even though No.2 sample has bigger particle which is not good for chemical absorption, the sorption ability is much better than that of pure $Li_4SiO_4$. The results show that the formation of eutectic substances can improve the sorption ability significantly.

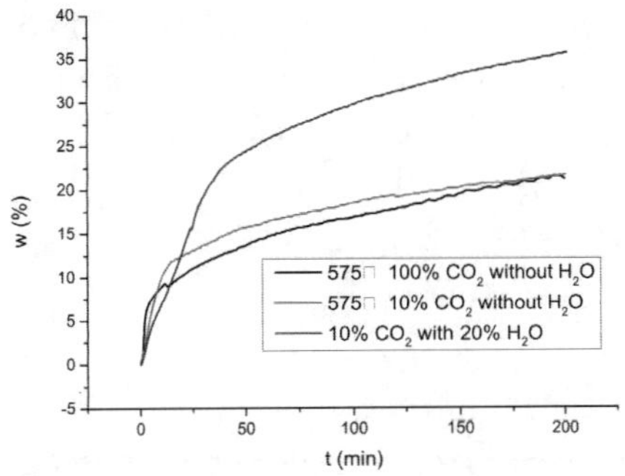


Fig.10 Isothermal curves of No.2 sample under different conditions

Fig.10 shows the Isothermal curves of No.2 sample under different conditions. For No.2 sample, the sorption ability maintains while the concentration of $CO_2$ decreases. While the water steam occurs, the sorption ability even improves with a weight gain of 20% with 30min. The improvement mainly displays in the extension of rapid reaction stage. When there is no steam, the rapid reaction stage ends when the weight gain is about 13% while 23% when there is steam. This can be attributed to the inhibition of the formation of double shell and the fact that steam can facilitate the mobility of $Li^+$ and $O^{2-}$.

## 4 Conclusion

(1) Doping K element into $Li_4SiO_4$ can form eutectic substance with the $Li_2CO_3$ produced by $Li_4SiO_4$ after $CO_2$ sorption thus facilitating the $CO_2$ sorption ability of $Li_4SiO_4$.

(2)It was comfirmed by calculation using Factsage and experiments results that the optimum doping ratio was $Li_2CO_3$:$SiO_2$:$K_2CO_3$=2.11:1:0.12. Then the rapid reaction stage of the adsorbent will be extended, and the maximum absorption capacity will be improved obviously.

(3)Under the same condition as in real SESRM process, the ability of $CO_2$ acceptors obtained maintains when the concentration of $CO_2$ decreases and the sorption ability even improves when steam occurs. The weight gain is more than 20% within 30min at 575℃, 10%$CO_2$ and 20% steam atmosphere.

## 5 References

[1] Li Li, Yuan Wenhui, Wei Chaohai. Adsorbents and adsorptive process for carbon dioxide at high temperature[J]. Chemical Industry and Engineering Progress, 2006, 25(8): 918-922

[2] Gauer C, Heschel W. Doped lithium orthosilicate for absorption of carbon dioxide[J]. Journal of Materials Science, 2006,41(8): 2405-2409

[3] M. Seggiani, M. Puccini, S. Vitolo. High-temperature and low concentration $CO_2$ sorption on $Li_4SiO_4$ based
sorbents: Study of the used silica and doping method effects[J]. International Journal of Greenhouse Gas Control 5 (2011): 741–748

[4] Mayra Y, Veliz-Enriquez, Gonzalo Gonzalez, Heriberto Pfeiffer. Synthesis and $CO_2$ capture evaluation of $Li_{2-x}K_xZrO_3$ solid solutions and crystal structure of a new lithium-potassium zirconate phase[J]. Journal of Solid State Chemistry,2007, 180: 2485–2492

[5] Robin W. Hughes, Dennis Lu, Edward J. Anthony, and Yinghai Wu. Improved long-term conversion of limestone-derived sorbents for in situ capture of $CO_2$ in a fuidized bed combustor[J]. Ind. Eng. Chem. Res, 2004, 43, 5529-5539

[6] Vasilije Manovic, Dennis Lu, Edward J. Anthony. Steam hydration of sorbents from a dual fluidized bed $CO_2$ looping cycle reactor[J]. Fuel, 87 (2008) 3344-3352

[7] Hendricus Th. J. Reijers, Sacha E. A. Valster-Schiermeier, Paul D. Cobden, and Ruud W. van den Brink. Hydrotalcite as $CO_2$ sorbent for sorption-enhanced steam reforming of methane[J]. Ind. Eng. Chem. Res. 2006, 45, 2522-2530

[8] Ping Sun, John R. Grace, C. Jim Lim, and Edward J. Anthony. Investigation of attempts to improve cyclic $CO_2$ capture by sorbent hydration and modification[J]. Ind. Eng. Chem. Res. 2008, 47, 2024-2032

[9] Aiyu Yan, Bin Liu, Yonglai Dong, Zhijian Tian, Dezheng Wang, Mojie Cheng. A temperature programmed desorption investigation on the interaction of $Ba_{0.5}Sr_{0.5}Co_{0.8}Fe_{0.2}O_{3\_\delta}$ perovskite oxides with CO2 in the absence and presence of $H_2O$ and $O_2$[J]. Applied Catalysis B: Environmental 80 (2008) 24–31

[10] Frank Zeman. Effect of steam hydration on performance of limesorbent for CO2 capture[J]. International Journal of Greenhoue Gas control, 2, ( 2008 ) 20 –209

[11] Esther Ochoa-Fernández, Tiejun Zhao, Ph.D., Magnus Rønning, De Chen. Effects of steam addition on the properties of high temperature ceramic $CO_2$ acceptors[J]. Journal of Environmental Engineering, 2009, 6: 397-403

# (350a) Treated Waste Water Source Heat Pump Design and Improvement

Huajun Yuan, Anum Raza, **Bo Song** and Sohail Murad, Chemical Engineering, University of Illinois at Chicago, Chicago, IL

A water source heat pump is a central heating/cooling system that pumps energy to or from water to provide heating or air conditioning. It is an efficient alternative to conventional heat, ventilation and air conditioning (HVAC) systems. Waste water treatment facilities provide easy access to large bodies of water which can be readily harvested. Such system generally lead to an almost 50% reduction in energy cost, and can therefore play an important role in maintaining and improving a sustainable ecosystem. The high flow rate and almost constant water temperature in such facilities make them especially attractive energy sources for heat pumps. We have collaborated with Metropolitan Water Reclamation Department of Greater Chicago (MWRDGC) to design and test different layouts of water source heat pump including open and close loops systems. The test data obtained will be used to increase the efficiency and lower maintenance costs, which will significantly reduce the cost of energy usage at the site. The test system has been installed at the Kirie Waste Water Treatment Facility in Des Plaines, IL. The system will in the future be further improved by using Phase Change Materials (PCMs) to take advantage of large latent heat of (about 250kJ/kg) PCMs to store energy during off-peak hours and release it during peak hours. This will reduce overdesign of the system and lower the initial capital cost. The entire process has also been simulated using the Aspen Plus simulation package and these simulations will allow us to further increasing the efficiency in the future as well. This project is a modest contribution to the development of more sustainable energy systems.

# (350b) Self-Heat Recuperation and Its Applications in Chemical Industries

**Yasuki Kansha**, Yuping Liu, Yui Kotani, Hiroyuki Mizuno, Masanori Ishizuka and Atsushi Tsutsumi, Institute of Industrial Science, The University of Tokyo, Tokyo, Japan

The combustion of fossil fuels for heating produces a large amount of carbon dioxide (CO2), which is the main contributor to the global greenhouse gas effect. Hence, the reduction of carbon dioxide (CO2) emission and the reduction of energy consumption for heating has nowadays become a very important issue in the efforts to suppress global warming. Heat recovery technology such as pinch technology, which exchanges heat between the hot and cold streams in a process, has been applied to thermal processes to reduce energy consumption. Although the net energy input seems to be reduced by using exhausted heat as the additional heat, the heat is also provided by the combustion of fossil fuels, leading to exergy destruction during energy conversion from chemical energy to heat.

Recently, Kuchonthara and Tsutsumi have proposed an energy and exergy recuperative integrated gasification power generation system and a design method for the system. Authors following on this concept and developed self-heat recuperation technology (SHRT) based on exergy recuperation. The most important characteristic of this technology is that the entire process stream heat can be recirculated into a process without any heat addition, leading to marked reduction of exergy and energy savings for the process.

In this research, the process design methodology based on self-heat recuperation is illustrated with the theoretical minimum energy required. Then, the energy required for processes based on self-heat recuperation and that for conventional processes with heat recovery in chemical industries are compared. From these results, the direction of energy saving process designs with fundamental developments is shown.

# (350c) Simulation of Self-Heat Recuperation Using Magnetocaloric Effect

**Yui Kotani**, Yasuki Kansha and Atsushi Tsutsumi, Institute of Industrial Science, The University of Tokyo, Tokyo, Japan

The reduction of carbon dioxide ($CO_2$) and the fossil fuel consumption has been one of our major concerns over past few decades. In chemical processes that involve heating, heat is mainly provided by fossil fuel combustion associated with great exergy loss. Recently, self-heat recuperation technology based on the exergy recuperative heat utilization principle has been developed, which can recirculate all heat of the process stream by providing work instead of heat generated by fuel combustion. By compressing gas at its highest temperature in the process for heat elevation, the energy consumption for heating can be reduced drastically compared to conventional heat-recovery technology.

A novel magnetic heat circulator for thermal process based on self-heat recuperation technology is proposed. In magnetic heat circulator, process heat is recirculated by magnetocaloric effect of ferromagnetic materials through cyclic magnetization and demagnetization. The ferromagnetic material is adiabatically magnetized/demagnetized at the highest/lowest temperature of the process to create a temperature difference required for heat exchange. All heat is recirculated inside the system without heat addition. The theoretical energy consumption of magnetic heat circulator was calculated in terms of the temperature-entropy diagram. The simulation result shows large potential for energy saving in thermal processes.

In this research, because the magnetic heat circulator poses a complex problem of heat transfer, magnetic field, and physical properties of ferromagnetic material, a numerical 1D model of magnetic heat circulator has been constructed. The magnetocaloric effect (MCE), varying magnetic field, thermal conduction and heat transfer has been implemented in the governing equation.

# (350d) A Novel Seawater Desalination Based On Self-Heat Recuperation

**Hiroyuki Mizuno**, Yasuki Kansha and Atsushi Tsutsumi, Institute of Industrial Science, The University of Tokyo, Tokyo, Japan

The demand for potable water has been increasing recently due to the rise in population and industrialization of developing country. There are two well-known desalination processes, thermal and membrane processes. Thermal process which separates potable water from seawater by evaporation can be divided into several types such as Multi Stage Flash and Multi Effect Desalination. Thermal desalination consumes large energy due to the heat requirement of water evaporation. But there are also merits; the contents of feed water do not strongly affect the desalination operation, and the salinity concentration of the product water is relatively low.

On the other hand, membrane processes such as Reverse Osmosis (RO) which separates portable water from brine water by semipermeable membrane is the most widely used desalination technique. Its reason is merely because it consumes less energy than conventional thermal process. Thus, most newly desalination plant is constructed based on RO process. However, when feed seawater contains large amount of dissolved solids and impurities, they damage the membrane. For this reason RO requires a greater degree of pretreatment than the thermal desalination process.

Summing up, thermal desalination does not need the greater degree of pretreatment but requires large amount of energy.

To overcome the problem, we proposed an energy efficient thermal desalination process based on self-heat recuperation. In a thermal process based on self-heat recuperation, both latent heat and sensible heat are recirculated into the process without any heat addition. As a result, energy requirement is drastically reduced.

In this study, the energy consumption of thermal process based on self-heat recuperation in the steady state has been calculated using a process simulator. We obtained some results; the energy consumption grows with increase of the minimum temperature difference in the heat exchangers, and increased quantities of production flow rate make the specific energy consumption smaller. The desalination based on self-heat recuperation consumes approximately 1/4 of the energy compared to the conventional thermal desalination process when the minimum internal temperature difference in heat exchangers is assumed to be 5 K and the percentage of the product in the feed stream is assumed as 30%.

By applying the self-heat recuperation technology to the thermal desalination process, we proposed an innovative thermal desalination process.

# (350e) Identifying Similar Environmental Objectives in Multi-Objective Optimization Via Clustering Methods

Diego Gabriel Oliva[1,2], **Gonzalo Guillén Gosálbez**[1] and Laureano Jiménez Esteller[1], (1)Department of Chemical Engineering, University Rovira i Virgili, Tarragona, Spain, (2)CONICET UTN, INGAR Instituto de Desarrollo y Diseño, Santa Fe, Argentina

Multi-objective optimization has recently emerged as a useful technique in environmental engineering, as it allows treating environmental aspects as additional objectives to be optimized rather than constraints on the system. One major limitation of these models (as applied to environmental engineering) is that they typically include a large number of objectives, which causes difficulties regarding the computation and visualization of the Pareto solutions.

Previous works by the authors focused on reducing the number of objectives in these problems using rigorous mixed-integer linear programs (MILP). In this piece of research, we have addressed a broader problem: given a multi-objective optimization model, we aim at grouping the environmental objectives into clusters, with the property that the minimization of any objective within a cluster will result in the minimization of the rest of objectives within the same cluster. Our final goal is to aid decision-makers by providing information on the structure of the problem, thereby offering them the opportunity of selecting the objectives to be minimized according to their preferences and the similarities between the environmental indicators of concern.

We present herein several theoretical and algorithmic developments in this area. We first provide a rigorous definition of clustering error in multi-objective optimization to quantify deviations from the Pareto structure of the problem that take place when optimizing in a reduced domain of objectives defined according to some clustering. We then introduce a new algorithm to perform such a clustering, and compare it with other traditional clustering methods available in the literature. Our clustering strategy is based on finding structural dominance relations between objectives, which are quantified according to different metrics

Finally, examples are taken from the literature to perform the comparison with other clustering methods. The main advantages of our method are its low running time and quality (i.e., error) of the clustering generated. The clusters identified provide in turn valuable insight for designers in multi-objective optimization.

# (350f) Holistic Multi-Phase Processing of Natural Materials

Aydin K Sunol, Chemical & Biomedical Engineering, University of South Florida, Tampa, FL and **Rosanna Ayala**, University of South Florida, Tampa, FL

A novel and green multi-fluid process for extraction and isolation of value added products from natural plants will be discussed. The process aims to remove components with different functional groups such as nutrients and anti-oxidants as well as utilize the residue beyond its heating value or energy conversion options. The solvents used are high pressure warm water and supercritical carbon dioxide while the plant used is acai berry. The extract fractionation beyond functional groups for isolation of chiral and isomers is part of extract removal component of the approach. The typical plant material residue options traditionally include its use as biomass. However, in this application residue is fermented for fruit wine. The pilot plant studies, product characterization, and process modeling using a flowsheet simulator illustrate the potential and flexibility of this dual green solvent process. The use of dual solvents multi-phase simultaneously where one is supercritical is a first in high pressure natural plant refining.

# (381a) Improvements in Bio-Butanol Separation by Adsorption: Adsorbent Screening, Kinetics and Equilibrium

Niloofar Abdehagh[1], F. Handan Tezel[2] and Jules Thibault[2], (1)Chemical and Biological Engineering Department, University of Ottawa, Ottawa, ON, Canada, (2)Department of Chemical Engineering, University of Ottawa, Ottawa, ON, Canada

The rising cost of crude oil combined with depletion of the resources, political instability in oil producing countries and environmental concerns are some of the reasons that have motivated different waves of interest for renewable and sustainable fuels. To achieve a greater use of biofuels, it is paramount to develop innovative techniques to convert biomass to biofuel as efficiently as possible in order to make these production processes more economically viable. Recent developments in biotechnology offer an avenue for bio-production of fuels from renewable and sustainable resources instead of petroleum-based fuels.

Bio-butanol offers several strong attributes and has advantages over other biofuels such as lower volatility, flammability and corrosiveness. The net heat of combustion (NHOC) for butanol is 29.2 MJ/L, which is 83% of gasoline NHOC (which is 32.5 MJ/L) whereas values for ethanol and methanol are 65 and 45% of gasoline NHOC, respectively. Butanol is a very hydrophobic compound with a solubility of 7.7 g/100 mL (at 20°C) in water,whereas lower alcohols (methanol, ethanol and n-propanol) are miscible in water. Also, butanol can be used in car engines without any modification. Butanol was initially produced from fermentation, but in the 1950s, butanol produced from petroleum via the oxo alcohol process became much cheaper and the chemical route remains the only economic source of butanol today. However, the resurgence of interest in the production of butanol from fermentation is being witnessed from research institutions and major players, including large multinational petroleum companies. However, the production of biobutanol via fermentation is facing significant engineering challenges due the very low final concentration, low yield and severe butanol toxicity to microorganisms. It is therefore important to find an efficient separation technique to recover butanol at the end of fermentation or during the fermentation to reduce the level of toxicity and prolong the fermentation in addition to making it an economically feasible process.

Adsorption is an energy-efficient process that can be used to achieve this objective. In selecting a suitable adsorbent for the adsorption process, many factors need to be considered such as : adsorption rate, adsorption capacity, ease of desorption, selectivity for the desired product and cost of the adsorbent. Activated carbon and zeolites are the most common adsorbents used in this process since they are hydrophobic materials having high adsorption capacity for butanol. In this study, two activated carbons (AC F600 and AC F400) with different particles sizes (0.55-0.75 and 1 mm) and two zeolites (ZSM-5 and NaY) with different $SiO_2/Al_2O_3$ ratios (80 and 1.8) were tested. Adsorbent screening was done based on the kinetic and equilibrium experiment results. In kinetic experiments, AC F400 and AC F600 showed much faster adsorption kinetics in comparison to zeolites., The results obtained from kinetic experiments for both activated carbons showed that the adsorption reaches equilibrium in approximately 100 minutes while for zeolites, the system did not yet reach equilibrium even after 240 min.

Results for equilibrium experiments showed that the highest adsorption capacity for butanol was obtained with AC F400. With an equilibrium butanol concentration of 15 g/L, the butanol adsorption capacity for AC 400 was 283 mg/g while butanol adsorption capacities for AC F600, ZSM-5 and NaY were 150, 155 and 116 mg/g, respectively. Based on these results, AC F400 was selected as the adsorbent to carry out further experiments. Due to the presence of other components such as acetone and ethanol in the fermentation broth in the butanol production process, it is essential to assess the impact of other components o[J1] n butanol adsorption capacity. To investigate the competitive effect of these components on butanol adsorption, ternary solutions of butanol-acetone-water and butanol-ethanol-water with different initial concentrations of acetone and ethanol (1, 5 and 10 g/L) were tested. In this series of experiments, results showed that the presence of ethanol has a negligible effect on butanol adsorption whereas the addition of acetone decreased the butanol adsorption capacity. The negative effect of acetone was more pronounced at higher acetone concentrations. For the 10 g/L butanol equilibrium concentration, in the presence of 1, 5 and 10 g/L initial acetone concentration, butanol adsorption capacity decreased from 258 mg/g to 239, 212 and 206 mg/g, respectively.

In this study, the selectivity of AC F400 for the three main fermentation species (butanol, acetone and ethanol) was investigated using binary solutions of butanol-water, acetone-water and ethanol-water separately. The results showed that AC F400 is more selective to butanol in comparison to the other components. At an equilibrium concentration of 10 g/L, adsorption capacity of AC F-400 for butanol was 258 mg/g, while for acetone and ethanol, the adsorption capacities at the same equilibrium concentration were 115 and 55 mg/g, respectively.

From all the results obtained in this investigation, activated carbon F400 has been chosen to be the best adsorbent for butanol separation due to its high affinity, fast kinetics and higher adsorption capacity.

---

[J1]Note that acetone is not an alcohol.

# (381b) 100% Conversion of Waste Brown Grease to Biodiesel and Syngas As Valuable Products for Heat and Power Cogeneration

**Shoucheng Du**[1], Fabio Poretti[2], Iman Noshadi[1], Richard Parnas[1] and George M. Bollas[1], (1)Department of Chemical, Materials & Biomolecular Engineering, University of Connecticut, Storrs, CT, (2)Department of Mechanical Engineering, University of Connecticut, Storrs, CT

Hydrocarbon-based biofuels derived from renewable, non-food carbon sources serve as a promising alternative to petroleum-derived fuels and are capable of mitigating greenhouse gas emissions. A sustainable biofuel strategy should utilize widely renewable biomass feedstocks, especially wastes, for which industries carry the high cost of disposal. Fuel production and power cogeneration from waste recovery is a sustainable and potentially profitable option to explore, addressing today's economic challenges of energy independence and waste management. A potentially interesting candidate in this scenario is brown grease. Brown grease is a mixture of high-value hydrocarbons, such as waste vegetable oil, animal fats and grease that is collected in water treatment plants and requires further treatment before disposal. However, brown grease has a higher level of contamination, compared to yellow grease, and cannot be used as animal feedstock or fertilizer. Additionally, brown grease is responsible for about 40% of all sewer overflows, causing back-ups and damage to pipe lines and for about 20,000,000 illnesses each year in the USA [1]. Therefore, special attention must be dedicated to its disposal and the cost associated. The main goal of this project is to explore the possibility of transforming 100% of the raw brown grease into biodiesel and synthesis gas that can be used for power generation. This work comprises a multi-disciplinary study of experimentation and simulation of optimal process options for the conversion of brown grease to energy and energy carriers.

In this study, heat treatment of the raw brown grease results in phase separation of the organic mixture. The light product, mainly characterized by a high free fatty acid (FFA) content (approximately 90%), is extracted and processed to obtain biodiesel. A two-step esterification process, using 5% (v/v) HCL and with a 9 to 1 methanol/oil molar ratio, produces liquid products with an acid value smaller than 0.5% FFA. The triglyceride is then transesterified using 2wt% KOH and a 6 to 1 of methanol/fats molar ratio, reaching higher ester content in the biodiesel. Gas Chromatography (Figure 1) shows that the biodiesel from the extracted brown grease oil not only satisfies the total and free glycerin percentage requirements imposed by the ASTM specifications, but also presents similar value in comparison with the biodiesel derived from Yellow Grease (Table 1), which makes it more competitive as brown grease is more abundant.

Table 1. Comparison of biodiesel produced from brown grease, yellow grease and ASTM standards in total and free glycerin

| Sample | Free glycerin | Total glycerin |
|---|---|---|
| Biodiesel from brown grease | 0.015 | 0.209 |
| Biodiesel from yellow grease | 0.022 | 0.158 |
| ASTM standard | <0.020% | <0.240% |

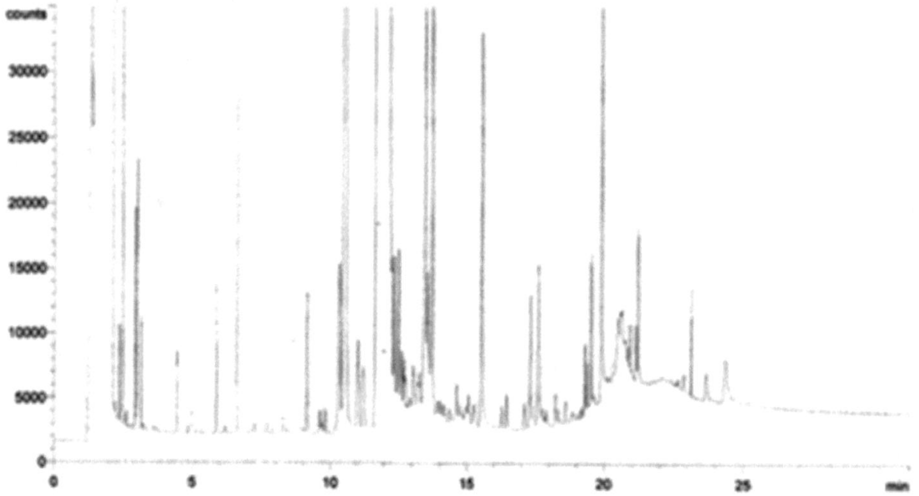


Figure 1. GC analysis of biodiesel produced from brown grease (Signal Count versus time)

The heavy product consists of a solid mixture, which is extracted and gasified to obtain synthesis gas. In Figure 2 thermogravimetric analysis (TGA) of the heavy brown grease product shows that this biosolid is gasifiable in its entirety. However the minor, almost negligible, residue of gasification contain inorganic compounds, of undefined environmental impact. Further analysis of the inorganics and their fate after gasification will be presented and discussed.

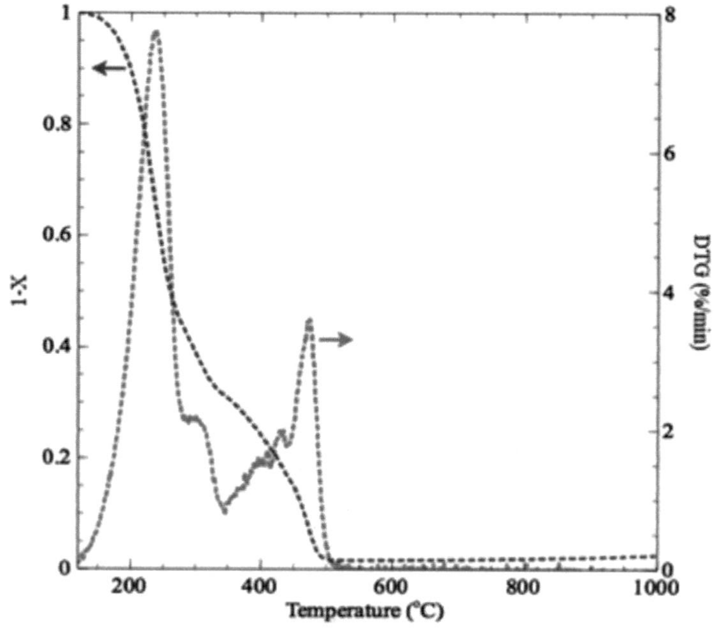


Figure 2. Gasification of biosolid residue from biodiesel production

TGA and gas product analysis (performed in gas chromatography/mass spectroscopy/Fourier transform Infrared, not shown here) show that the heavy product of brown grease is rich in hydrogen; thus, a valuable feedstock. Elemental analysis (Table 2) shows that the carbon to hydrogen ratio of the brown grease heavy fraction biosolid is lower than pine sawdust, which confirms its value as a gasification feedstock.

Table 2. Elemental analyses of biosolid from biodiesel production, biomass and coal (mol%)

| Feedstock | N | C | H | S | C:H |
|---|---|---|---|---|---|
| Biosolid | 0.0263 | 3.2483 | 6.4692 | 0.0157 | 0.5021 |
| Sawdust | 0.0261 | 4.4880 | 6.0236 | 0.0201 | 0.7451 |
| Coal | 0.1007 | 5.9767 | 5.0600 | 0.0881 | 1.1812 |

Due to the relatively low available quantity of the brown grease heavy fraction (as opposed to the efficiency increase of large scale gasification), co-gasification of the brown grease residue with fossil fuels, such as coal, is investigated. Co-gasification of coal and brown grease residues is capable of increasing the power generation plant throughput; thus, increasing its overall efficiency [2]. On the other hand, co-gasification reduces the consumption of coal in power generation plants resulting in economic and environmental benefits. Experimental results of co-gasification of biosolids and coal will be compared with the more widely studied co-gasification of woody biomass and coal [3].

On the basis of experimental results, process simulation of an integrated brown grease biorefinery is performed in Aspen Plus. The objective of the simulation study is to identify the cost and profitability of an integrated brown grease refinery with power co-generation, focusing also on the benefits of capturing CO2 and the overall reduction of greenhouse gas emissions. The energy penalties for drying, O2 separation, H2S, SOx, NOx, and CO2 separation are estimated for various co-gasification scenarios. A brief description of the process flow sheet is as follows. Raw brown grease is fed to the system and the first operation consists of heating the mixture in order to evaporate all the containing water and to phase separate the substance in the two layers. The oil at the top layer is extracted and sent to the biodiesel production hierarchy, while the heavier biosolids are taken to the syngas production hierarchy for gasification. In the biodiesel production hierarchy the liquid animal fat enters the pretreatment unit where impurities are filtered and removed. Cleaned animal fat enters an esterification unit where the oil is homogenously mixed at a constant temperature of 60°C with Methanol (MeOH) and Hydrochloric acid (HCl), which serves as a catalyst. The reaction reduces the concentration of the FFA. Then the stream is washed and dried. A second esterification process follows, to convert all the FFA remaining into Methyl ester. The stream is washed and dried again to remove remaining methanol and acid catalyst. The processed flow then enters the Transesterification unit where a base catalyst, Potassium hydroxide (KOH), is used to convert the triglycerides into biodiesel. This reaction also produces glycerol, which can be separated and sold as a chemical. Finally, crude biodiesel enters the post-processing unit where it is refined and tested for quality standards.

In the syngas production hierarchy the dewatered biosolids enter the preparation unit where they are mixed with coal, which has previously passed a process of sizing (crushing and screening) and drying. Air is separated mainly in Nitrogen and Oxygen in the Air Separation Unit (ASU) through cryogenic distillations. The mixed feedstock, together with the stream of almost pure Oxygen, enters the gasifier. The melted ash is collected at the bottom of the reactor and disposed. Then the syngas is cleaned from H2S and CO2 in the Acid Gas Cleaning Unit (Rectisol Process). Captured CO2 is sent to compression and storage, and the H2S is converted to solid S in the Claus Unit [3]. Finally the syngas is burned in a combined cycle, where heat and electricity are produced. The hot gas, after expansion in the gas turbine, pass through a Heat Recovery Steam Generator (HRSG) where superheated, high pressure steam is produced which can be used for district heating and/or expanded in a steam turbine for additional power production.

The experimental analysis integrated with the ASPEN model allows to examine the feasibility of recovering 100% of the raw brown grease that would otherwise have to be disposed at a high cost, due to its contamination as a water treatment waste. The opportunity of transforming it into valuable energy products opens access to new markets, while the sustainability of the overall process and the improved carbon life cycles of this process, lead to its reassessment from an unwanted waste to a precious resource.

Acknowledgment

This work was supported in part by the Faculty Large Grant of the University of Connecticut and in part by the EAGLES (Engineers as Global Leaders for Energy Sustainability) Atlantis Fellowship.

References:

[1] Tyson, K. S. Brown Grease Feedstocks for Biodiesel, National Renewable Energy Laboratory, 2002.

[2] Saw, W.; McKinnon, H.; Gilmour, I.; Pang, S. Fuel. 2012, 93, 473-478.

[3] Valla, L.; Du, S.; Bollas, G. M. 2012 AICHE Annual Meeting, October 2012, Pittsburgh, PA, USA.

# (381c) Techno-Economic Analysis of Mixed Alcohol Production Through Hydrogenation of Volatile Fatty Acids Derived From Seaweed

Peyman Fasahati, Chemical Engineering, Pukyong National University, Busan, South Korea and **J. Jay Liu**, Department of Chemical Engineering, Pukyong National University, Busan, South Korea

Recently, there has been growing interest regarding use of seaweeds as biomass feedstocks. On bioethanol production from seaweeds, there are two main obstacles that make it hardly applicable in industrial scale. First, high seaweed price greatly effects the economy of bioethanol production from seaweeds. Developing mass cultivation sites and enhancing seaweeds cultivation procedures would be necessary steps to address this problem. Secondly, ethanol production from seaweeds has relatively low yield. Research on finding new enzyme, bacteria, or yeast, which are capable of fermenting a wider range of sugars produced by seaweed saccharification, can be the only solution. An alternative way to produce bioethanol would be finding new substitute processes that produce ethanol with higher conversion and lower final ethanol price.

Volatile fatty acids (VFAs) can be easily produced from seaweeds. They have higher conversion yields in comparison with those of direct ethanol production. VFAs can easily be converted to ethanol by hydrogenation. For these reasons, this study develops a techno-economic model to survey economy of VFAs production from seaweeds followed by hydrogenation to produce bioethanol.

Our previous study on direct bioethanol production from seaweed showed that the maximum dry seaweed price (MDSP) should be lower than 160 US$ per ton to allow ethanol production economically viable. This study also calculates the MDSP for this process and survey possible opportunities to improve the economy of ethanol production.

The whole process was designed and simulated in Aspen plus V7.3 and then sent to aspen process economic analyzer for economical evaluations. Plant scale-up effects on the economy of plant were also examined in detail.

**Flex Fuel Polygeneration: Optimizing Cost, Sustainability, and Resiliency**

*Matthew Kieffer, Tristan Brown, Robert C. Brown*

Iowa State University
Center for Sustainable Environmental Technologies

Flex Fuel Polygeneration (FFPG) is a new technology that employs multiple energy sources and produces multiple energy carriers to construct optimal energy plant designs in terms of cost, sustainability, and resiliency. This technology allows for adjustments to be made to market fluctuations and improvements to the energy efficiency and environmental performance of power plants. The diversification of energy pathways also increases resiliency to contingencies such as earth quakes, flooding, reduced winds, or other events that may reduce the availability of energy sources. In this study we develop two FFPG plants that employ combinations of natural gas, biomass, wind, and the use of anaerobic digestion, fuel cells, different types of turbines, and Fischer-Tropsch synthesis. To obtain FFPG plant performance results, we combine techno-economic analyses available in literature of existing energy plants to reflect the FFPG plant designs. By doing high level analysis we can compare costs, products, and emissions of theoretical plants to plants that are currently in operation. What we have found are preliminary results regarding the internal rates of return. If the internal rates of return prove to be positive, we will continue to design and perform high level analysis on additional FFPG plants.

# (381e) Feed Flexible Process Optimization Model for Region Specific Sustainable Biorefining

**Sumesh Sukumara**[1], Jeffrey R. Seay[2], Dr. Fazleena Badurdeen[3], William Faulkner[4] and Joseph Amundson[4], (1)Chemical and Materials Engineering, University of Kentucky, Lexington, KY, (2)Department of Chemical and Materials Engineering, University of Kentucky, Paducah, KY, (3)Institute for Sustainable Manufacturing, University of Kentucky, Lexington, KY, (4)University of Kentucky, Lexington, KY

Biofuels in recent years has emerged as one of the most promising alternatives to replace fossil fuel. Researchers around the world are working on developing new processes and improve existing ones to convert biomass into various marketable products. Integrated biorefining is a possibility which combines a slate of processes involving various feedstock to produce transportation fuel, electric power and other useful chemical products. Also, due to the seasonal and regional variability of feedstock, the overall supply chain has to be looked at carefully. Thus, a multidisciplinary economic model needs to be developed which incorporates various small models and allows each to exchange information. According to the existing literature and preliminary experiments performed its evident that transportation and processing cost share a significant portion of the overall cost of an integrated biorefinery. This novel contribution aims to show how a feed flexible process optimization model can be developed and interact with the overall region specific optimization model.

Processing cost in a biorefinery is mainly due to the utilities and raw materials involved. The model developed applies various concepts of chemical engineering to minimize the utilities and maximize the profitability for any given process in a region. Various modules of the Aspen Engineering Suite are combined to achieve the required objective. The process optimization model to be described consists of two major processes to produce transportation fuel and power as end products. These models are first developed considering all the possible mass and heat streams involved to get various products. A thermal pinch analysis is performed to find the best possible heat integration practically possible for a process. Based on the optimum heat exchanger network the process simulations are changed accordingly and finally the capital costs are estimated for the process. The process optimization model developed transfers its data with a Visual Basic interface to the other models which eventually estimates the overall profitability of the process. This model will be validated using the Jackson Purchase region of Western Kentucky, which is not only rich in coal, lignocellulosic biomass and crop residue but also has large number of poultry farms, as a case study. The robust model developed can be run for various combinations of feedstock possible. These process simulation results coupled with the overall supply chain model can predict some critical conclusions like profitability, optimum feed ratio, utility requirement and also some sensitive process variable that effect the overall supply chain of the biorefining process. This innovative linking of various optimization models will help develop a large scale multivariable model for biorefining which will include all possible feedstocks for a given region and provide the option to choose from various available processes in the model based on the cost, demand and environmental regulations in a region. Hence the multiple feedstock process optimization model developed supports the overall supply chain model based on which investors can check the economics of a process and decide if an investment is worthy in a given region.

# (381f) Sustainable Design of Energy Supply Chains From P-Graphs: Cost, Ecological Footprint, and Emergy

Leisha Vance[1], Istvan Heckl[2], Botond Bertok[3], **Heriberto Cabezas**[1] and Ferenc Friedler[4], (1)Office of Research and Development, U.S. Environmental Protection Agency, Cincinnati, OH, (2)Department Computer Science and Systems Technology, University of Pannonia, Veszprem, Hungary, (3)Department of Computer Science, University of Veszprém, Veszprém, Hungary, (4)Department of Computer Science and Systems Technology, University of Pannonia, Veszprém, Hungary

A collaboration consisting of the Office of Research and Development (ORD) of the U.S. EPA and the research group led by the founders of the P-graph framework at the University of Pannoniahas developed a methodology for designing sustainable supply chains based on an optimization using the p-graph framework constrained by integrated sustainability indicators and engineering costs. The result was a powerful methodology for designing cost-effective and environmentally sustainable supply chains. We illustrated the methodology with a prototype supply chain designed to produce heat and electricity for a generic district in Hungary. Possible sources of heat and electricity included electricity from the Hungarian grid, and heat and electricity generated from natural gas, corn, corn silage, grass silage, or wood or some combination of these sources. Twenty-one different supply chains, each capable of producing 18 TJ per year of heat and 7.2 TJ per year of electricity were found. Each supply chain was ranked according to cost, and assessed environmental impacts using the ecological footprint (representing land use burden), and emergy (representing energy resource burden). Decisively feasible supply chains were found with cost variations of +2% to -17% compared to "business as usual" scenarios i.e. using only natural gas and electricity from the Hungarian grid. For these supply chains, the sustainability profile as represented by the ecological footprint varied from +8% to -78%, and the emergy results ranged from -54% to -93%. Both comparisons were done in contrast to the conventional natural gas/electricity from the Hungarian grid. Most importantly, it appeared feasibly possible to design supply chains for heat and electricity generation which were both cheaper and more sustainable than the supply chain currently in use.

# (385a) Assessing the Potential for Large-Scale Utilization of Fossil-Fuel Derived CO2

Jared Ciferno, U.S. Department of Energy, National Energy Technology Laboratory, Pittsburgh, PA and **John J. Marano**, JM Energy Consulting, Inc., Gibsonia, PA

Barring the imposition of a carbon tax (or equivalent cap-and-trade program) and corresponding international agreements, it is highly unlikely that carbon capture and storage/sequestration (CCS) will be implemented in any meaningful way in the near future. This is due to inherent inefficiency and higher costs associated with this option for mitigating climate change. Unfortunately, the science of climate change suggests that the world community must begin to act sooner rather than later. From an Industrial Ecology viewpoint, a preferable approach to carbon storage in perpetuity would be to re-use the carbon associated with the fossil-fuel derived $CO_2$ already being omitted. However, options for directly utilizing $CO_2$ are quite limited, with enhanced oil recovery (EOR) being the only physical use of $CO_2$ of any appreciable size. The concept of converting $CO_2$ to useful products has been explored in the past, but has fallen from favor in many policy circles due to the lack of a clear vision of how re-utilization of fossil carbon could be implemented. Given current in-action, it is worth re-visiting this concept.

Energetically, it is clear that the large-scale conversion of $CO_2$ to useful products requires a source of external energy that is not fossil-fuel derived; otherwise, the conversion process becomes itself a source of $CO_2$ emissions. Sources for the required energy could be direct solar or nuclear energy, or electricity generated from these primary energy resources. Obviously, a source of $CO_2$ is also required; however, all sources are of equal value. Atmospheric and mobile emissions are of little value due to the problems associated with collecting and concentrating them. Whereas, $CO_2$ from stationary point sources, and in particular large point sources such as coal and natural gas-fired power plants, are the "low hanging fruit" for implementing carbon re-use. Finally, useful products from $CO_2$ must be identified. It is suggested here that instead of pursuing a piecemeal approach consisting of an assortment of finished chemicals and material goods, a holistic approach be taken. Currently, the only existing commodity market that is big enough to absorb the large scale re-utilization of $CO_2$ lies in the petroleum sector. Thus, the basis for any carbon re-utilization platform should be the production of a petroleum substitute, a "synthetic" oil that may be refined and converted into a multitude of finished products.

This paper first compares and contrasts carbon storage with carbon utilization, and then moves on to quantify the scale of any carbon re-utilization effort; as well as, $CO_2$ life-cycle limitations of re-use. After all, in this approach much of the synthetic oil, like petroleum, will end up being (re-) combusted as an end-use transportation fuel. Thus, carbon re-use can only be considered as an interim strategy for cost effectively transitioning away from fossil fuels. We also briefly touch on the implications of large scale carbon re-use for CCS R&D, since the requirements for $CO_2$ capture, transport and short-term storage will be quite different than what is anticipated for long-term $CO_2$ sequestration. Finally, we briefly discuss the possible economic implications of this strategy. A follow-on paper at this conference identifies and analyzes what the present and best future options for producing a $CO_2$-derived synthetic oil might be.

# (385b) Some Perspectives On the Use of Solar and Other Renewable Energy to Facilitate Carbon Management

**Robert Wegeng**, Pacific Northwest National Lab, Richland, WA

The recent and ongoing development of economical solar and wind energy systems presents a number of new options for carbon management. The purpose of this presentation will be to discuss how engineered solar thermochemical processes might be used to reduce carbon emissions from fossil-fuel powerplants and gas-to-liquids plants and how concentrated solar energy might be used to capture carbon from industrial emissions or the atmosphere for sequestration or use. Perspectives will include economics and Second Law efficiencies. Particular attention will be applied to technologies and strategies that might be applied in the U.S. Pacific Northwest.

# (385c) Carbon Capture and Storage Directly From Flue Gas: The Effects of Chemical and Biological Catalysts On Magnesium Carbonate Formation

**Edward J. Swanson**, Chemical Engineering, Columbia University, New York, NY, Patrick V. Brady, Sandia National Laboratories, Albuquerque, NM and Ah-Hyung Alissa Park, Earth and Environmental Engineering & Chemical Engineering, Columbia University, New York, NY

The direct adsorption of $CO_2$ from flue gas and precipitation of mineral carbonates presents a transformative alternative to carbon capture and storage (CCS) schemes involving conventional post-combustion capture of $CO_2$ and its geologic storage. $CO_2$ capture processes generally involve energy intensive solvent regeneration and CO2 compression steps. Absorbing the carbon into an aqueous system and precipitating it as a thermodynamically stable solid carbonate in a single technology avoids both of these energy intensive steps in CO2 capture, while providing inherent storage of CO2. However, in order for this novel CCS technology to be viable, the kinetics of both CO2 absorption and mineral dissolution need to be sufficiently fast, particularly at the low partial pressure of CO2 commonly found in flue gas. In this study, both chemical (i.e., catechol and oxalate) and biological (i.e., carbonic anhydrase) catalysts were used to enhance the rates of CO2 hydration and serpentine dissolution. While these catalysts showed significant improvement over natural carbon mineralization kinetics, it was also important to investigate the effect of this catalytic system on the final formation of magnesium carbonate in terms of its chemistry and morphology. Experiments were performed in a lab-scale three-phase fluidized bed reactor, and optimum reaction conditions for carbonate precipitation were investigated.

# (385d) Environmentally Benign Processes for Absorption and Chemical Fixation of CO2

**An-Hua Liu**, Ran Ma and Liang-Nian He Sr., State Key Laboratory of Elemento-Organic Chemistry, Nankai University, Tianjin, China

Carbon dioxide as the main green house gas is widely blamed to pay most of the responsibility for global warming. Hence, controlling $CO_2$ emission and further reducing its accumulation emerges to be an urgent and challenging research topic. $CO_2$ capture and storage/sequestration (CCS) to control its emission as well as chemical utilization of the captured $CO_2$ (CCU) has been considered as the rational technique to reduce or mitigate the so-called green-house effect. However, being the sustainable carbon source, $CO_2$ as a safe, cheap and abundant $C_1$ building block in organic synthesis also comes into light [1-3]. So chemical fixation of $CO_2$ into industrially useful materials and high value-added products turns to be one of the most attractive methods in recent years.

Amine-functionalized $CO_2$ absorbents have been intensively investigated in numerous studies due to the easy preparation, high $CO_2$ affinity and robust reversibility, whereas the intrinsic nature of 2:1 absorption stoichiometry could restrain their further development for pursuing of superior capture efficiency. In present work, such inherent disadvantage was circumvented by employing modified amino acid salts in polyethylene glycol, affording approximative 1:1 ratio $CO_2$ absorption. The used absorbent can be readily reverted and recycled without loss of its capture capacity at least 5 turns. The structure of captured absorbent species was identified by means of NMR and in situ IR spectroscopy, as well as computational calculations.

Meanwhile, using commercial available and recyclable basic ion-exchange resins, one kind of polystyryl supported tertiary amines, an efficient fixation of $CO_2$ by aziridines under mild conditions were achieved, leading to formation of 5-aryl-2-oxazolidinone with excellent regio-selectivities. Notably, neither solvent nor any additive was required, and the catalyst can be recovered by simply filtration and directly reused at least five times without significant loss of catalytic activity and selectivity [4]. The present protocol was successfully applied to cycloaddition reactions of epoxides/propargyl amines with $CO_2/CS_2$. In general, this solvent-free process thus represents environmentally friendly catalytic conversion of $CO_2$ into value-added chemicals and has great potential to be applied in various continuous flow reactors in industry.

**References**

[1] Aresta, M.; Dibenedetto, A. *Dalton Trans*. 2007, 28, 2975.

[2] Yang, Z.-Z.; He, L.-N.; Zhao, Y.-N. *Energy Environ. Sci*., 2011, 4, 3971-3975.

[3] He, L.-N.; Wang, J.-Q.; Wang, J.-L. Carbon dioxide chemistry: examples and challenges in chemical utilization of carbon dioxide, *Pure Appl. Chem*. 2009, 81(11), 2069-2080.

[4] Liu, A.-H.; He, L.-N.; Peng, S.-Y.; et al. *Sci. China Chem.,* 2010, 53(7), 1578-1585.

# (385e) Process Ecoefficiency of Methanol and Dmc Production From CO2: A Comparative Analysis

Camila Ribeiro[1], José L. Medeiros[2], **Ofélia Araújo**[2], Paulo Coutinho[3] and Rita M.B. Alves[4], (1)Escola de Química (H2CIN), UFRJ - Universidade Federal do Rio de Janeiro, Rio de Janeiro, Brazil, (2)Escola de Química (H2CIN), UFRJ - Universidade Federal do Rio de Janeiro, Rio de Janeiro, Brazil, (3)management of innovation, Braskem, São Paulo, Brazil, (4)Braskem, São Paulo, Brazil

**Process Ecoefficiency of Methanol and DMC Production from $CO_2$: A comparative analysis**

Camila F. Ribeiro[1], Ofélia de Q. F. Araújo[1], José Luiz de Medeiros[1], Rita[2], Paulo Coutinho[2]

[1]Departament of Chemical Engineering, Escola de Química da Universidade Federal do Rio de Janeiro. Sala E-201, Bloco E, Centro de Tecnologia, Ilha do Fundão, CEP: 21949-900 Rio de Janeiro, Brazil.

E-mail adress: camila@eq.ufrj.br

Environmental performance of products and processes has gained importance as the industry is trying to minimize its environmental impacts. Of special concern to society, are climate effects associated to the rise of $CO_2$ atmospheric concentration. To mitigate emissions effects, processes are searched that use $CO_2$ as raw material. It is worth noting that an emission mitigating process should necessarily be a commodity as large production scale are required. With this orientation, methanol and DMC are promising alternative products: the first is a very attractive additive to diesel, and it blends with gasoline to yield a high octane fuel with low Reid vapor pressure. Methanol, on the other hand, is considered as a promising energy alternative besides displaying a wide range of industrial applications. However, in order to claim that a given technology is clean, its ecoefficiency should be addressed with quantitative sustainability indicators. In this sense, the present work selects a set of environmental metrics along with safety and economical (CAPEX and OPEX) indicators. A comparative analysis is presented for two potentially green processes: (i) methanol synthesis; and (ii) DMC synthesis. Furthermore, the later synthesis is approached in two alternatives, a direct and an indirect reaction path. In the methodology used, process simulation and optimization are employed to compare process performances at maximum yields.

# (385f) High Efficiency Direct Electrochemical Conversion of CO2 to Liquid Fuel

**Jingjie Wu**[1], Frank Risalvato[2], Fusheng Ke[2], Perry Pellechia[3] and Xiao-Dong Zhou[2], (1)Chemical Engineering, University of South Carolina, Columbia, SC, (2)Department of Chemical Engineering, University of South Carolina, Columbia, SC, (3)Department of Chemistry and Biochemistry, University of South Carolina, Columbia, SC

In this paper, we will discuss a design of an electrochemical cell for $CO_2$ reduction implementing a proton exchange membrane (PEM) by adding a liquid buffer layer between the membrane and the cathode electrode. This configuration enables production of liquid fuels from $CO_2$ on Sn gas diffusion electrode at high efficiencies. Specifically, the formate was produced from the electrochemical reduction of $CO_2$ with an overpotential lower than 0.2 V. At such a low overpotential, the Faradaic efficiency and energetic efficiency of formate production with a buffer layer of 0.1 M $KHCO_3$ reaches 78.5% and 66.4%, respectively. Moreover, a maximum Faradaic efficiency of ~95% can be achieved. The production rate of formate increases with increasing cell potential to approximately 10 μmol $min^{-1}$ $cm^{-2}$ at a cell potential of -3.0 V. In contrast, hydrogen evolution reaction dominates at the cathode side in the PEM electrochemical cell without a buffer layer.

The enhancement of $CO_2$ reduction in the PEM electrochemical cell with a buffer layer is attributed to the optimization of $H^+$ and $CO_2$ concentration at the reactions sites in the cathode. The effects of $H^+$ concentration on the Faradaic efficiency were confirmed by the results of $CO_2$ reduction with a liquid buffer layer with different pH values. As a result, the Faradaic efficiency towards formation of formic acid can be further improved by optimizing the buffer layer thickness and subsequently tuning the $H^+$ concentration at the cathode surface. The optimized thickness of the buffer layer was found ~ 3 mm. The buffer layer thickness dependency of $H^+$ concentration at the cathode electrode will be reported.

# (385g) Electrochemical Conversion of Carbon Dioxide to Useful Products

Arun Agarwal[1], Shan Guan[2], Davion Hill[2], Edward Rode[2] and Narasi Sridhar[2], (1)Research & Innovation, Det Norske Veritas (DNV), Dublin, OH, (2)Det Norske Veritas (DNV)

Utilization of captured $CO_2$ provides an advantageous alternative especially in locations where $CO_2$ sequestration in geologic media is not feasible. Making high value chemical products from $CO_2$, although not sufficient to convert a significant portion of the GHG emissions, provides an opportunity of extensive production of green chemicals when combined with renewable sources to feed the energy requirements. At DNV, we are developing the technology to economically convert $CO_2$ to formate salts/ formic acid using an electrochemical process. Tin is the preferred electrocatalyst due to high selectivity of $CO_2$ reduction to formate ion and high overpotential for cathodic hydrogen evolution. Gaseous $CO_2$ is continuously purged into the catholyte (2M KCl) through a porous tin electrode on which it is cathodically reduced. This cathode chamber is separated from the anodic chamber using a selective ion exchange membrane for allowing positive ion transport but not allowing the product (formate) ions to get oxidized on the anode. Oxygen evolution reaction occurs on a mixed metal oxide anode in either acidic or alkaline anolyte solutions to complete the electrochemical reaction. Electrode development in terms of high surface area of catalyst to match existing commercial electrochemical processes and reduce capital costs related to reactors will be presented. A process was developed to electroplate tin on high surface area carbon fiber substrate. Electroplating solution and substrate properties were evaluated for achieving maximum current densities at optimal product selectivity. Another focus area is understanding and optimizing the process chemistry. Selection of anolyte and catholytes, along with the applied potential regime, affects product selectivity, electrode degradation, expenditure due to chemicals consumption and the ability to reach high enough product concentrations before feasibly employing separation processes. In this presentation, a technical and economic analysis of the aforementioned factors affecting the commercial feasibility will be presented and evaluated as a guide for performance criteria of the proposed technology.

# (414a) Technological and Economic Analysis of Adsorption Processes for Direct Capture of CO2 From Air

**Ambarish R. Kulkarni**, Stephanie A. Didas, Christopher W. Jones and David S. Sholl, School of Chemical & Biomolecular Engineering, Georgia Institute of Technology, Atlanta, GA

Direct capture of $CO_2$ from air is a concept that can be potentially used to account for $CO_2$ emissions from mobile and disperse sources that are inaccessible to flue gas capture. Initially, we study equilibrium process models to determine the viability of adsorption-based air capture technologies using an amino-modified silica adsorbent, TRI-PE-MCM-41, and a structured monolithic contactor unit. Using the purity of $CO_2$ and annual product throughput as metrics for comparing process performance; we identify some of the operational parameters, adsorbent characteristics and sources of parasitic energy losses that have a significant effect on the performance of the process. Specifically, we will discuss a process with a daily throughput of ~1.1 $tCO_2$ at 88.5 $\%CO_2$ purity using standard shipping container sized air capture modules. By accounting for the available sources of energy (low pressure steam and electricity) and geographic location, we obtain a net operating cost for air capture of $CO_2$ to be ~ \$100/$tCO_2$. Recent results from more detailed adsorption models and experiments will also be presented. Finally, potential strategies for reducing the energy and monetary cost of these processes will be discussed. Our analysis supports continued work to establish the technological and economic feasibility of adsorption-based air capture.

# (414b) Dynamic Modeling and Control of a Solid-Sorbent $CO_2$ Capture Process with Two-Stage Bubbling Fluidized Bed Adsorber-Reactor

**Srinivasarao Modekurti**, Department of Chemical Engineering, West Virginia University, Morgantown, WV, Debangsu Bhattacharyya, Department of Chemical Engg., West Virginia University, Morgantown, WV and Stephen E. Zitney, AVESTAR Center, National Energy Technology Laboratory, Morgantown, WV

Solid-sorbent-based $CO_2$ capture processes have strong potential for reducing the overall energy penalty for post-combustion capture from the flue gas of a conventional pulverized coal power plant. However, the commercial success of this technology is contingent upon it operating over a wide range of capture rates, transient events, malfunctions, and disturbances, as well as under uncertainties. To study these operational aspects, a dynamic model of a solid-sorbent-based $CO_2$ capture process has been developed.

In this work, a one-dimensional (1D), non-isothermal, dynamic model of a two-stage bubbling fluidized bed (BFB) adsorber-reactor system with overflow-type weir configuration has been developed in Aspen Custom Modeler (ACM). The physical and chemical properties of the sorbent used in this study are based on a sorbent (32D) developed at National Energy Technology Laboratory (NETL). Each BFB is divided into bubble, emulsion, and cloud-wake regions with the assumptions that the bubble region is free of solids while both gas and solid phases coexist in the emulsion and cloud-wake regions. The BFB dynamic model includes 1D partial differential equations (PDEs) for mass and energy balances, along with comprehensive reaction kinetics. In addition to the two BFB models, the adsorber-reactor system includes 1D PDE-based dynamic models of the downcomer and outlet hopper, as well as models of distributors, control valves, and other pressure-drop devices. Consistent boundary and initial conditions are considered for simulating the dynamic model. Equipment items are sized and appropriate heat transfer options, wherever needed, are provided. Finally, a valid pressure-flow network is developed and a lower-level control system is designed.

Using ACM, the transient responses of various process variables such as flue gas and sorbent temperatures, overall $CO_2$ capture, level of solids in the downcomer and hopper have been studied by simulating typical disturbances such as change in the temperature, flowrate, and composition of the flue gas. To maintain the overall $CO_2$ capture at a desired level in face of the typical disturbances, two control strategies were considered–a proportional-integral-derivative (PID)-based feedback control strategy and a feedforward-augmented feedback control strategy. Dynamic simulation results show that both the strategies result in unacceptable overshoot/undershoot and a long settling time. To improve the control system performance, a linear model predictive controller (LMPC) is designed. In summary, the overall results illustrate how optimizing the operation and control of carbon capture systems can have a significant impact on the extent and the rate at which commercial-scale capture processes will be scaled-up, deployed, and used in the years to come.

# (414c) CO2 Separation Using Mg(OH)2 Solutions – Dissolution Kinetics and Vapor-Liquid-Solid Equilibrium Studies of CO2- Mg(OH)2 System

Hari Krishna Bharadwaj[1], Joo-Youp Lee[1] and Tim.C Keener[2], (1)Chemical Engineering Program, University of Cincinnati, Cincinnati, OH, (2)Environmental Engineering Program, University of Cincinnati, Cincinnati, OH

Coal-fired power plants are responsible for a considerable amount of total anthropogenic $CO_2$ emissions throughout the world. $CO_2$ from flue gases can be absorbed by $Mg(OH)_2$ slurry at 52 °C in a column followed by a regeneration step in a stripper operated at 65 °C.

One of the most important steps in the $CO_2$ scrubbing process is the rate at which magnesium hydroxide dissolves from the solid part of the slurry and becomes available for reaction with absorbed $CO_2$. The dissolution rate kinetics of $Mg(OH)_2$ was studied using a pH stat and the dissolution rate was determined based on acid consumption required to maintain a preset pH value. The effects of pH, temperature, mixing and particle size on the dissolution rates of $Mg(OH)_2$ will be presented. A dissolution rate kinetic expression obtained under different operating conditions will also be presented.

Knowledge of Vapor liquid solid equilibrium (VLSE) of $CO_2$ in $Mg(OH)_2$ is essential for the proper design and operation of absorption and desorption equipment. The VLSE data for the $Mg(OH)_2$ – $CO_2$ system under various operating conditions of pressure, temperature, and concentration of $Mg(OH)_2$ will be presented. A suitable model for fitting the VLSE data will also presented.

# (414d) CO2 Capture Using Particulate Silica Aerogel Immobilized with Tetraethylenepentamine (TEPA)

Jerry YS Lin, Robert Pfeffer and **Nick Linneen**, Chemical Engineering, Arizona State University, Tempe, AZ

$CO_2$ capture using particulate silica aerogel immobilized with tetraethylenepentamine (TEPA)

Nick Linneen (presenter), Jerry YS Lin, Robert Pfeffer

School for Engineering of Matter, Transport, and Energy

Arizona State University

Tempe AZ 85289

Abstract:

Post-combustion capture of CO2 using solid amine-based sorbents is attracting significant attention due to their high selectively, reduced corrosive properties, and low energy regeneration requirements. Research has shown that sorbents that have large pore diameter and pore volume generally achieve the highest CO2 adsorption capacities. This is a result of the support's greater potential to retain large amounts of amine, while also allowing effective CO2 transport within the sorbent. We have developed a novel sorbent using particulate aerogel as a support. Aerogel, also known as "solid smoke," is a high porosity, high surface area, silica-based material that has very large pore volumes and pore diameters compared to other supports. Therefore aerogels can retain large amounts of amine while also maintaining some mesoporosity for effective CO2 transport. By impregnating tetraethylenepentamine (TEPA), which is an ultra-low molecular weight linear polyethylenimine (PEI), into commercially available 10 micron size aerogel (Cabot Corp. Nanogel), we have obtained an adsorption capacity as high as 6.1mmol/g which is very competitive relative to other CO2 sorbents. The aerogel-amine based sorbent was also found to achieve fast kinetics reaching 90% of its equilibrium capacity within the first 10min and also maintaining its adsorption capacity after 10 adsorption/desorption cycles.

# (414e) Development of High Performance Carbon-Based Adsorbents for CO2 Adsorption

**Placidus B. Amama**, RXBT/UDRI, Air Force Research Laboratory, Wright-Patterson AFB, OH and Patrick J. Shamberger, RXBT, Air Force Research Laboratory, Wright-Patterson AFB, OH

Thermal management of high-power systems is a pressing challenge for many USAF platforms. These systems require rapid absorption and ultimately rejection of low-quality (low temperature) heat. Activated carbon is capable of adsorbing pressurized $CO_2$ gas with a high heat of adsorption, offering the possibility of rapidly absorbing and rejecting heat at low temperatures. The activated carbon adsorbent has to meet important performance criteria for high efficiency of the system. The performance criteria include high adsorption capacity, high adsorption kinetics, excellent stability and mechanical properties, and high thermal conductivity. For activated carbon adsorbents to meet all of these criteria, the $CO_2$ adsorption capacity and the adsorption-desorption kinetics must be improved. The adsorption of $CO_2$ on activated carbon can be significantly enhanced by the introduction of basic nitrogen functionalities on the surface. In this work, surface modification of a series of activated carbon adsorbents via gas phase amination under various conditions has been performed. The surface area, pore structure and pore volume were characterized using $N_2$ physisorption while the types of chemical functionalities present on the surface after amination were characterized by X-ray photoelectron spectroscopy (XPS), Fourier transform infrared spectroscopy (FTIR), and temperature programmed desorption (TPD). $CO_2$ adsorption on the modified activated carbon samples have been measured gravimetrically using a high-pressure TGA.

# (442a) Sustainable Applications Through the Use of Reversible Ionic Liquids

**Steven R. Saunders**[1,2], Kyle M. Flack[1,3], Emily C. Nixon[1,3], Amy L. Rohan[1,2], Jackson R. Switzer[1,2], Elizabeth J. Biddinger[1,2], Pamela Pollet[1,3], Charles L. Liotta[1,2,3] and Charles A. Eckert[1,2,3], (1)Specialty Separations Center, Georgia Institute of Technology, Atlanta, GA, (2)School of Chemical and Biomolecular Engineering, Georgia Institute of Technology, Atlanta, GA, (3)School of Chemistry and Biochemistry, Georgia Institute of Technology, Atlanta, GA

Reversible ionic liquids (RevILs) are a novel class of switchable solvents that are being investigated for use in a wide variety of applications. Switchable solvents make use of an external physical stimulus to induce a chemical change which causes a drastic, abrupt change in the properties of the solvent. The RevILs discussed herein are formed when $CO_2$ is introduced to a silylamine (e.g., 3-(aminopropyl) triethylsilane) to form an ammonium-carbamate ion pair. These RevILs can be returned to their original molecular (i.e., non-ionic) form when heated or sparged with an inert gas. Through adjustments in the structure of the silylamine and the degree of conversion of the molecular form to the ionic form the properties (e.g., viscosity, density, polarity, etc.) of the solvent medium can be drastically tuned. This tuning ability makes these RevILs ideal solvent systems for a number of applications.

Two applications of particular interest where RevILs have made a significant impact in terms of reducing emissions and waste are carbon capture of $CO_2$-rich gas streams and nanoparticle synthesis. These RevILs provide several advantages over traditional $CO_2$ capture solvents (e.g., monoethanolamine) for recovery of $CO_2$ from, for example, coal fired power plant flue gas. The primary advantage involves a dual capture mechanism: (1) chemical absorption and (2) physical absorption which provides for the opportunity to reach similar recoveries as traditional solvents with lower energy requirements. Traditional synthesis routes for catalytically active nanoparticles can make use of harsh reactants and produce significant amounts of waste. Nanoparticles with relatively narrow size distribution and an average size less than 10 nm can be produced using these RevILs in a process akin to a reverse-micelle synthesis with the ability to recycle the RevIL These two applications offer a focus on waste reduction and the environmental sustainability of the processes.

# (442b) Application of Ionic Liquid for the Efficient and Energy-Saving Synthesis of Ethylene Glycol

**Jian Sun Sr.**, State Key Laboratory of Multiphase Complex System, Institute of Process Engineering, Chinese Academy of Sciences, Beijing, China, Weiguo Cheng, State Key Laboratory of Multiphase Complex System,, Institute of Process Engineering, Chinese Academy of Sciences, Beijing, China, Jinquan Wang Sr., Key Laboratory of Green Process and Engineering, Institute of Process Engineering, Chinese Academy of Sciences, Beijing, China, Junping Zhang, Institute of Process Engineering, Chinese Academy of Sciences, Beijing, China and Suojiang Zhang, 1.State Key Laboratory of Multiphase Complex System, 2.Key Lab of Green Process and Engineering, Institute of Process Engineering, Chinese Academy of Sciences, Beijing, China

**Application of Ionic Liquid for the Efficient and Energy-Saving Synthesis of Ethylene Glycol**

Jian Sun, Weiguo Cheng, Jinquan Wang, Junping Zhang, Suojiang Zhang*

*Institute of Process Engineering, Chinese Academy of Sciences,100190, Beijing, PR China*

**Corresponding author: sjzhang@home.ipe.ac.cn*

Ethylene glycol (EG) is chemical of importance for synthetic fibers, resins, and antifreeze. The most widely used commercial process to manufacture EG is the direct non-catalytic liquid-phase hydration of ethylene oxide (EO). A large excess of water (20-25 mol water/mol EO) must be used to obtain a high selectivity of EG in the hydration reaction mixture, which consumes a large steam input for products separations. Even so, an achieved typical product mixture is only approximately 90% EG. Therefore, the need to increase EG selectivity but reduce energy cost has spurred the development of alternative routes and the corresponding catalysts.

Recently, the manufacture of EG from EO via ethylene carbonate (EC) intermediate has been regarded as one of the most efficient routes to replace the current method. However, the catalysts still suffer from low activity, harsh reaction conditions and water/air sensitivity. In addition, the requirement of EC separation might be a response to an unsatisfactory compatibility of the two-step reaction catalysts.

Since ionic liquids (ILs) have shown an excellent performance for the synthesis of industrial chemicals substituting for traditional catalysts, herein, we designed and screened binary IL catalysts for the new process, between which one would be active towards the cycloaddition reaction and another could be effective for the hydration process. Effect of various parameters on the new process was studied in detail. It was found that in the presence of water, 1) EO conversion could be enhanced obviously, 2) the molar ratio of components in catalyst had a great effect on the selectivities of EC, EG and diethylene glycol (DEG), and 3) the catalyst was hydrothermally stable and could be reused over 10 times without loss of its activity and selectivity. Based on the results, an economic comparison of the direct hydration process and the new process was investigated by aspen-plus simulation. The total process avoids EC separation, and represents a simple, energy-saving, and cost-effective route to EG with high product quality, as well as easy catalyst recycling.

**Keywords: ionic liquid, ethylene glycol, carbon dioxide, synthesis, energy-saving**

**Acknowledgments**

This work was supported byNational Basic Research Program of China (2009CB219901), and National Science Fund of China (21006117 and 20936005).

# (442c) A Comparative Study On Selective Alkylation of Naphthalene Catalyzed by Various Lewis Acidic Ionic Liquids

**Ming Zhang**, Zhiqiang He and Yun Fang, School of Chemical & Material Engineering, Jiangnan University, Wuxi, China

A series of double long-chain alkylnaphthalene were synthesized from naphthalene and long-chain α-alkenes catalyzed by various Lewis acidic ionic liquids in this paper. The Lewis acidic ionic liquids could be luckily used as catalyst instead of traditional acidic catalysts even zeolites in the process of alkylation with high activity because traditional acidic catalysts have several disadvantages, such as corrosion, low selectivity, and environmental pollution since difficultly recoved. More than that, the alkylated products and unreacted reactants could be isolated easily by decantation from the immiscible ionic liquid phase. In this paper, various Lewis acidic ionic liquids composed of $AlCl_3$-$Et_3NHCl$, $ZnCl_2$-$Et_3NHCl$ or $FeCl_3$-$Et_3NHCl$ were tested and the influences of type and dosage of catalyst, cationic-anionic proportion of ionic liquids, molar ratio of naphthalene to alkenes, reaction temperature, and reaction time were investigated. Excellent conversion of reactants and different selectivity of the alkylated product were achieved, and finally the catalyst could be recycled for several times without any obvious loss of its catalytic activity.

# Use of Deep Eutectic Ionic Liquid for Biodiesel Preparation from Base-Catalyzed Transesterification of Rapeseed Oil

*Wei Huang* [a], *Shaokun Tang,* *[a] *Hua Zhao* [b]

[a] *Key Laboratory for Green Chemical Technology of Ministry of Education, School of Chemical Engineering & Technology, Tianjin University, Tianjin 300072, China*

[b] *Chemistry Program, Savannah State University, Savannah, GA 31404, USA*

**Corresponding author. Tel: +86-22-27408578; E-mail address:* shktang@tju.edu.cn

As a renewable and biodegradable fuel, biodiesel has been attracting research attention for the last decade in the face of diminishing oil supplies. To overcome hurdles of the conventional reaction and purification processes for biodiesel preparation, a number of new approaches have been actively pursued including new solid catalysts, novel reaction process, and co-solvents in recent years. However, these new methods are associated with new problems such as lower product yield, higher energy consumption, higher costs and so on.

In our work, a deep eutectic ionic liquid (DEIL), choline chloride and glycerol (1:2, molar ratio), was prepared and used in the transesterification reaction of rapeseed oil to biodiesel catalyzed by NaOH. Response surface methodology (RSM) and Box-Behnken Design (BBD) were employed to examine different reaction parameters such as methanol/oil molar ratio, catalyst concentration and DEIL concentration, and optimize the reaction conditions for achieving the maximum yield of fatty acid methyl ester (FAME).

Our data have suggested that the FAME yield of up to 88% could be obtained at the optimum conditions; this yield was higher than that of the reaction under the same condition parameters without the addition of DEIL (74%). A statistical analysis of variance (ANOVA) indicates that a second-order polynomial model appears to represent the actual relationship between the response and the significant variables, with a very low probability value (0.0001) and a high satisfying determination coefficient ($R^2$=0.9724).

The addition of DEIL to the transesterification reaction improved the FAME yield and enabled an easy and thorough separation of glycerol from the reaction mixture during and after reaction. This study exemplifies a cost-effective and environmental benign approach to biodiesel production by using the DEIL as a co-solvent for the transesterification reaction.

**Keywords**: deep eutectic ionic liquid, biodiesel, transesterification, base catalyst, response surface methodology

**Acknowledgment**

This project was financially supported by the National Natural Science Foundation of China (21176186).

# (442e) Ionic Liquid Design and Techno-Economic Evaluation for Energy Efficient Extractive Recovery of Bio-Based Glycols

André B. de Haan, Boelo Schuur and Lesly Garcia-Chavez, Chemical Engineering, Eindhoven University of Technology, Eindhoven, Netherlands

Important bulk chemicals like glycols can be produced from renewable sources through fermentation processes. An important challenge is the energy efficient product recovery from diluted fermentation broths. Low molecular weight diols, such as Mono Ethylene Glycol, Propylene Glycols and Butane Diols are particularly difficult to separate due to their high water affinity. Their recovery by distillation or multiple effect evaporation is associated with high energy consumption. Liquid-Liquid extraction can be a promising alternative since it can be much more energy efficient. However, for conventional solvents the distribution coefficients are generally insufficient to achieve efficient extraction at low broth concentrations. For this reason advanced fluid extractants like ionic liquids are needed that improve the glycol distribution.

In this work COSMO-RS is used to support ionic liquid tailoring to optimize the glycol distribution coefficient (D) and selectivity (S) by employing the sigma electron profile. As a result the glycol distribution coefficients were increased up to >100 times that of conventional solvents. For the best performing ionic liquid tetraoctyl ammonium 2-methyl-1-napthoate liquid-liquid equilibrium data were determined and thermodynamic models constructed for various glycols. These thermodynamic models were used to develop conceptual process designs that demonstrated energy savings >50% compared to the current state-of-the-art three effect evaporation technology. Energy integration appeared crucial to achieve these energy savings in the extraction process. In spite of these significant energy savings the techno-economic evaluation demonstrated that the final economic feasibility of the extraction routes strongly depends on the capital expenditure for the respective glycols, which in turn is strongly dependent on the final glycol distribution coefficient achieved.

# (442f) Dissolution of Activated Sludge Biomass in Imidazolium-Based Ionic Liquids

**Patrisha J. Pham**, Rafael Hernandez and Todd French, Dave C. Swalm School of Chemical Engineering, Mississippi State University, Mississippi State, MS

Activated sludge biomass (ASB), the solid produced during biological treatment of wastewater contains a network of polysaccharides and proteins collectively known as extracellular polymeric substance (EPS). The nature of ASB predicts that it may behave similarly with cellulose upon dissolution in ionic liquids (IL). Cellulose dissolution in ionic liquids is well-known and studied. In this work, ASB was dissolved in various ionic liquids such as 1-butyl-3-methylimidazolium chloride ([BMIM][Cl]) and 1-allyl-3-methylimidazolium chloride ([AMIM][Cl]), both of which are recognized to dissolve cellulose effectively. The biomass was recovered after precipitation with water. The regenerated ASB was characterized by Thermogravimetric Analysis (TGA). The bulk structure was observed through Scanning Electron Microscopy (SEM). Factors such as the nature of the ionic liquid, ASB homogeneity and temperature effects (microwave irradiation versus direct heating) on dissolution were investigated. Establishing successful dissolution parameters of ASB in imidazolium-based IL's will open doors for new ASB utilization i.e. sugar source for biofuels production.

# (442g) Liquid-Liquid Extraction of Alcohols From Aqueous Solutions Using Ionic Liquids

**Alexander Nann**[1], Christoph Held[1], Martin Stoffers[2], Andrzej Górak[2] and Gabriele Sadowski[1], (1)Department of Biochemical and Chemical Engineering, Laboratory of Thermodynamics, TU Dortmund University, D-44227 Dortmund, Germany, (2)Department of Biochemical and Chemical Engineering, Laboratory of Fluid Separations, TU Dortmund University, D-44227 Dortmund, Germany

The production of alcohols and fuels via fermentation has been suggested as an alternative to the fossil-fuel based routes in recent years. Liquid-liquid extraction of organic compounds from dilute aqueous solutions using ionic liquids is a promising unit operation to replace energy- and cost-intensive separation techniques, e.g. distillation. The selection of appropriate ionic liquids (ILs) applied in chemical and separation processes requires reliable thermodynamic data. Therefore, systematic investigations of pure-component and mixture properties for systems containing ILs are essential. This work investigates the phase behavior of binary and ternary alcoholic solutions containing ILs as well as the extraction of 1-butanol from dilute aqueous solutions.

In a first step, physical and thermodynamic properties of pure IL and binary IL/alcohol mixtures of four ILs consisting of two different anions and two different cations were measured. The experimental investigations were carried out using vapor-pressure osmometry (VPO), gas-liquid chromatography (GLC), and a vibrating-tube densimeter at different temperatures and concentrations.

In a second step, pure-component IL model parameters for PC-SAFT were obtained from fitting to pure-IL densities and binary IL/alcohol data. The ILs were regarded as non-dissociated as well as non-aggregated. This approach allows for a quantitative modeling of densities and alcohol activity coefficients over a broad concentration and temperature range.

Subsequently liquid-liquid equilibrium (LLE) phase diagrams for ternary water/alcohol/IL systems were determined experimentally. It was found that all considered ionic liquids seem to be promising agents for liquid-liquid extraction of the considered alcohols from dilute aqueous solutions. Using the IL model parameters based on only pure and binary component data, PC-SAFT was able to predict the experimental LLEs of these systems precisely.

Finally, with knowledge of the liquid-liquid phase behaviour a multistage liquid-liquid extraction is investigated. Therefore an equilibrium model was implemented to Aspen Custom Modeler®. Important process parameters and the IL demand were identified by varying the number of theoretical stages, butanol fraction and phase ratio.

# (480a) Recovery of Hydrocarbons From Froth Treatment Tailings of Oil Sand Bitumen Processing

Yuming Xu, CanmetENERGY, Natural Resources Canada, Devon, AB, Canada and Jianmin Kan, Total E&P Canada, Calgary, AB, Canada

Bitumen is separated from surface-mined oil sand ore in a warm water extraction process that produces bitumen froth typically containing 60 wt% bitumen, 30 wt% water, and 10 wt% mineral solids. The water and solids are then removed from the froth by solvent dilution followed by gravitational settling. If light alkanes are used as the solvent in the froth treatment process, some of the asphaltenes in the bitumen are precipitated and form aggregates with solids and water droplets. These aggregates quickly separate from the diluted bitumen phase leaving very clean diluted bitumen. After solvent recovery, the asphaltenes, water, and mineral solids are discharged as froth treatment tailings.

The asphaltenes contained in the tailings constitute valuable hydrocarbons that, if recovered, could be useable as coker feed or paving material. We conducted investigations of the feasibility of recovering asphaltenes from the tailings. Two methods were tested – *solvent extraction* and *agglomeration*. In *solvent extraction*, aromatic solvent is mixed with froth treatment tailings to dissolve the asphaltenes followed by centrifugation to remove the mineral solids and water. The experimental results demonstrated that more than 90% of the asphaltenes in the tailings can be recovered. The recovered asphaltenes contain only a small fraction of mineral solids and might be usable as coker feed. In *agglomeration*, the tailings are agitated at elevated temperature (80°C) causing the asphaltene particles to form large agglomerates that separate from the tailings to give an asphaltene-rich sediment phase and tailings that are almost asphaltene-free. The agglomerates still contain significant fractions of mineral solids and water and require further treatment.

# (480c) Molecular Dynamics Simulations of Fly-Ash Based Geopolymers: Conformational Changes At High Temperatures

**Fernando A. Soto**[1], Ancy Kunjumon[1], Kunal Kupwade-Patil[1], Erez Allouche[1] and Daniela S. Mainardi[2], (1)Louisiana Tech University, Ruston, LA, (2)Chemical engineering, Louisiana Tech University, Ruston, LA

**Molecular Dynamics Simulations of Fly-Ash based Geopolymers: Conformational Changes at High Temperatures**

Fernando Soto, Ancy Kunjumon, Kunal Kupwade-Patil, Erez N. Allouche and Daniela S. Mainardi

Geopolymers are an emerging class of cementitious materials, which offer a sustainable, low energy consuming; low carbon footprint and 100% substitute to Portland cement as a cementitious binder in engineering applications. In addition, geopolymers offer an intriguing mixture of superior physical properties, extensive environmental benefits and strong prospects for commercial implementation in high performance refractory applications. The objective of this study is to gain an atomistic understanding of the structural evolution of geopolymeric gels. In particular, the ultimate goal is to develop insights as to how the molecular structure of geopolymers change with time at elevated temperatures. Density Functional Theory (DFT), Molecular Mechanics (MM) and Molecular Dynamics (MD) approaches are employed to model and simulate geopolymeric behaviors.

In this work, two models of hydrated sodium aluminosilicates (NASH) polymers were considered. NASH models with Si:Al ratio of 1:1 and 2:1 were simulated at selected temperatures ranging from 273 K to 4273 K. The study provided information on stability, and structural and transport properties of geopolymers. Furthermore, Gibbs Free-Energy barriers were calculated to gain an in-depth knowledge of reaction rates that are essential for understanding the geopolymerization mechanism.

# (480d) Effects of Farm Wastes Co-Digestion On Microbial Community, Methane Production, and Fiber Quality As Biorefining Feedstock

**Rui Chen**[1], Fan Yang[2], Zhengbo Yue[3], James MacLellan[1], Terence L. Marsh[2], Yan Liu[1] and Wei Liao[1], (1)Biosystems and Agricultural Engineering, Michigan State University, East Lansing, MI, (2)Microbiology and Molecular Genetics, Michigan State University, East Lansing, MI, (3)Resources and Environmental Engineering, Hefei University of Technology, Hefei, China

The performance of anaerobic digestion is highly correlated to the compositions of its substrate and microbes. A lab-scale anaerobic digestion process of a mixture of dairy manure and corn stover in continuous stirred-tank reactor (CSTR) was performed to investigate the influence of additional organic carbon on methane biogas production, lignocellulose-degrading efficiency, microbial community, and fiber quality of solid digestate for ethanol production. Highest biogas production of 3,158 mL day$^{-1}$ was observed at hydraulic retention time (HRT) of 40 days in the digester fed on manure and corn stover. Biogas productivity was closely correlated with the populations of polymer degrading bacteria *Bacteroidetes* (particularly *Unclassified Bacteroidetes*) and *Clostridia,* as well as methanogenic archaea such as *Methanosarcina* and *Methanosaeta*. The solid digestate (AD fiber) from co-digestion process showed better conversion efficiency in the enzymatic hydrolysis. Results also indicated that anaerobic co-digestion of corn stover and dairy manure significantly enhanced the energy balance of an integrated anaerobic digestion/bioethanol production process.

# (480e) Continuous Hydrothermal Liquefaction of Livestock Waste Using a Plug-Flow Reactor

Mitchell James Minarick, Agricultural and Biological Engineering, University of Illinois at Urbana-Champaign, Urbana, IL

Direct liquefaction of biomass is a technology that has shown good potential for producing oil from biomass, especially for wet biomass sources like algae and livestock waste. Previous research at the University of Illinois at Urbana-Champaign proved that up to 70% of swine manure volatile solids could be converted into oil while reducing the manure's chemical oxygen demand by up to 75%. A continuous stirred-tank reactor design was implemented and produced similar conversion rates to the batch process. Algae conversion rates are similarly high for batch processes, and integrated systems that utilize a manure source to provide nutrients for enhanced algae growth could simultaneously clean the waste stream while producing algae for oil production.

However, the high pressures required for the direct liquefaction process preclude the scale-up of continuous stirred tank reactor systems. A plug-flow reactor was designed and tested to explore the feasibility of a scaled-up continuous reactor for the liquefaction of the biomass from this integrated bioenergy system.

# (480f) Aeration Induced pH Elevation of Anaerobically Digested Wastewater for Enhanced Ammonia Stripping and Recovery

**Quanbao Zhao**, Department of Biological Systems Engineering, Washington State University, Pullman, WA and **Craig Frear**, Washington State University, Pullman, WA

During mesophilic or thermophilic anaerobic digestion, a considerable fraction of organic nitrogen is mineralized to ionic ammonia, producing an effluent with higher concentrations of ammonia nitrogen as well as elevated temperatures, both of which can be useful in dowstream processing of ammonia via ammonia stripping. Unfortunately, release of the ammonia also requires a suitable increase in pH well above that normally existing in digested wastewaters. Chemical dosing for pH elevations at or above 10 is possible but economically problematic. Our research has shown that limited, non-biological aeration at laboratory conditions of 0.5-1 L/min of the digested wastewater (dairy and poultry manure) at varying temperature regimes (35, 55, 70C), results in intriguing chemical shifts strongly related to the high levels of carbon dioxide, bicarbonates and carbonates present in digested effluent. These chemical shifts, ultimately release carbon dioxide and raise the pH of the effluent to levels near 10, which combined with elevated operating temperatures (from waste engine heat in commercial scenario), can lead to 70-90% shift from ionic to free, gaseous form of ammonia and subsequent recovery of the ammonia through acid contact. Length and extent of aeration time, percent recovery of ammonia, and final pH attained is influenced by wastewater source and in particular its levels of suspended solids, chemical composition, buffering capacity, and level of ammonia binding to struvite or other salt compounds. The resultiing stripped effluent shows an important enhancement in degree of suspended solids (TSS) and total phosphorus (TP) settling, with 50 and 80% reductions, respectively. This oral presentation is a summary of laboratory results with dairy manure wastewater, with particular emphasis on the chemical relationships and equilibrium shifts associated with the aeration process and its subsequent release of gases. Additional information will also summarize scale-up and demonstration of the process with industrial partners.

# (480g) Aspen Simulation of Mobile Biomass to Liquid Facility – the Beauti-Fuel Concept

Xinying Liu[1], Bilal Patel[1], Diane Hildebrandt[2] and David Glasser[2], (1)Centre of Material and Process Synthesis, University of the Witwatersrand, Johannesburg, South Africa, (2)Centre of Material and Process Synthesis (COMPS), University of the Witwatersrand, Johannesburg, South Africa

The BeauTi-fueL facility is a mobile biomass to liquid facility, where the key equipment of the biomass to liquid process is placed into a commercial container and can be transported to a biomass site for liquid fuel production. This facility can be used to process the biomass such as agriculture and forestry waste as well as mulniciple solid waste, to convert them into energy products and can be useful for energy needed remote area. In the facility, biomass is fed into a plasma gasifier. The gas existing the gasifier is then cooled and cleaned before sent to the Fischer Tropsch synthesis unit. Liquid hydrocarbons formed in the Fischer Tropsch Synthesis is knocked out in the cooling and separation system and the tail gas with light hydrocarbons and unreacted CO and H2 is sent to a gas engine to generate electricity. The electricity generated is to be used for the process itself.

The reason why a plasma gasifier was selected was because of its compactness and low tar formation because of the high temperature achieved. The gasification chamber of the plasma gasifier with the needed capacity is small and can be fit into the container, where the size of the fixed bed gasifier and fluidized gasifier are all too large for this situation. The disadvantage of using a plasma gasifier is it utilizes a high quality energy form, which is electricity, to provide energy needed for the gasification process. This raises the question of the overall energy efficiency of the whole process.

An aspen simulation is conducted on this process concept at the claimed capacity of 1 barrel per day liquid hydrocarbon production. Process conditions such as air feeding rate, air enrichment, plasma gasification operation temperature and pressure, the Fischer Tropsch synthesis operation pressure, syngas conversion, hydrocarbon selectivity and the gas engine efficiency are studied. It is found that with the optimized process condition and the right biomass feed, only 1 dry ton per day biomass fed is needed for 1 barrel per day production.

# (481a) Integration of Post-Combustion Capture with Compression and Liquefaction Process of CO2

**Yeong Su Jeong**[1], Ung Lee[1], Seeyub Yang[1], Youngsub Lim[1], Chi Seob Lee[2] and Chonghun Han[1], (1)School of Chemical and Biological Engineering, Seoul National University, Seoul, South Korea, (2)KEPCO E&C Company, Yongin, South Korea

Along with the worldwide efforts to reduce the greenhouse gases, Carbon Capture and Storage/Sequestration (CCS) has emerged as one of the most promising technologies in South Korea. Capture process is known to occupy 70 to 80 percent of the overall energy demands in CCS technology, and consequently, most of up-to-date research has been dedicated to this aspect by developing innovative $CO_2$ absorbents or modifying the process itself to reduce the regeneration energy of the absorbent. Each constituent of the technology can then be integrated with one another in order to mitigate the overall energy consumption. However, there is no possible candidate for $CO_2$ storage site in Korea at this time, so the capture process must be integrated with an intermediate compression process instead which is always required to transport $CO_2$from capture plant to the storage site.

In this study, a process simulation including both post-combustion capture of $CO_2$ using MEA absorbent and multi-stage compression was performed using Aspen Plus®. Different physical property methods were selected for each process to allow more stable and accurate simulation. [1] It was assumed that the compressed (liquefied) $CO_2$ will be transported via $CO_2$ carrier, which requires appropriate specifications of $CO_2$ for ship loading. [2] In addition, 4-stage compression system with intercooling was adopted where both cold and hot utilities are available from compression as well as isenthalpic expansion. A quantitative analysis of how these utilities can be used was carried out so that the energy required for capture as well as the overall process was reduced.

This research was supported by a grant from the LNG Plant R&D Center funded by the Ministry of Land, Transportation and Maritime Affairs (MLTM) of the Korean government.

References:

[1] Van Wagener, D.H., Rochelle, G.T., 2011. Stripper configurations for $CO_2$capture by aqueous monoethanolamine, Chem. Eng. Res. Des. 89, 1639-1646.

[2] Aspelund A., Jordal, K., 2007. Gas Conditioning–The interface between $CO_2$ capture and transport, Int. J. Greenhouse Gas Control, 1(3), 343-354.

# (481b) A Hybrid Model-Based Framework for the HAZOP Analysis of Geological Storage of Carbon Dioxide

**Greeshma Gadikota**[1], Babji Srinivasan[1], Ah-Hyung Alissa Park[2] and Venkat Venkatasubramanian[1], (1)Chemical Engineering, Columbia University, New York, NY, (2)Earth and Environmental Engineering & Chemical Engineering, Columbia University, New York, NY

With ever increasing energy demand, carbon capture, utilization and storage (CCUS) has become one of the greatest challenges facing the global community since carbonaceous fuels are still the most abundant and inexpensive sources of energy. CCUS is a complex multistep process which involves (i) capture of $CO_2$ from flue gas generated in coal-fired power plants (ii) transportation of $CO_2$ using long pipelines to the storage site, and (iii) injection of $CO_2$ into the geological formations. Each of these steps pose a number of potential risks such as impurities associated with $CO_2$ capture, leakage in pipelines, corrosion in pipelines, leakage and/or migration of $CO_2$ in the geological formations due to faults and fractures, contamination of freshwater aquifers etc.Therefore, it is essential to perform a systematic HAZOP analysis for these processes before implementation of any of the CCUS technologies.

In this work, a hybrid model-based framework is proposed to perform HAZOP analysis for CCUS. First, detailed signed directed graph-based qualitative models are constructed for various components with specific emphasis on $CO_2$ storage. Further, dynamic simulations of the various tasks involved in the complex CCUS process are performed using quantitative models. These quantitative dynamic simulations with an emphasis on $CO_2$ storage, supplement the overall graph theoretical model built using process-knowledge. Together, these models are utilized to understand the behavior of the entire complex process under various conditions and more importantly, to evaluate the risks present in the overall system. Furthermore, these graph theoretic models with dynamic simulations are explored to perform fault detection, diagnosis and monitoring in the overall CCUS process with emphasis on $CO_2$ storage in geological media.

# (481c) Impact of Alkalinity Sources On the Life-Cycle Energy Efficiency of Mineral Carbonation Technologies

**Abby Kirchofer**, Earth, Energy, and Environmental Sciences, Stanford University, Stanford, CA, Adam Brandt, Energy Resources Engineering, Stanford University, Stanford, CA, Sam Krevor, Petroleum Engineering, Imperial College, London, England, Valentina Prigiobbe, Petroleum and Geosystems Engineering, University of Texas at Austin, Austin, TX and Jennifer Wilcox, Department of Energy Resources Engineering, Stanford University, Stanford, CA

Impact of alkalinity sources on the life-cycle energy efficiency of mineral carbonation technologies

Abby Kirchofer[1], Adam Brandt[2], Sam Krevor[3], Valentina Prigiobbe[4], and Jennifer Wilcox[2]

[1] Earth, Energy, and Environmental Sciences, Stanford University

[2]Energy Resources Engineering, Stanford University

[3]Petroleum Engineering, Imperial College

[4]Petroleum and Geosystems Engineering, University of Texas at Austin

Mineral carbonation is a Carbon Capture and Storage (CSS) technology where gaseous $CO_2$ is reacted with alkaline materials (such as silicate minerals and alkaline industrial wastes) and converted into stable and environmentally benign carbonate minerals (Metz et al., 2005). Previous studies on mineral carbonation have principally focused on identifying process conditions optimal for enhancing the chemical rates of reactions. No study, however, has yet performed a scheme that optimizes these conditions with respect to engineering and economic consideration for the goal of producing a viable carbonation process.

Here, we present a holistic, transparent life cycle assessment model of aqueous mineral carbonation built using a hybrid process model and economic input-output life cycle assessment approach. We compared the energy efficiency and the net $CO_2$ storage potential of various mineral carbonation processes based on different feedstock material and process schemes on a consistent basis by determining the energy and material balance of each implementation (Kirchofer et al., 2011). In particular, we evaluated the net $CO_2$ storage potential of aqueous mineral carbonation for serpentine, olivine, cement kiln dust, fly ash, and steel slag across a range of reaction conditions and process parameters. A preliminary systematic investigation of the tradeoffs inherent in mineral carbonation processes was conducted and guidelines for the optimization of the life-cycle energy efficiency are provided.

The LCA model allows for the evaluation of the tradeoffs between different reaction enhancement processes while considering the larger lifecycle impacts on energy use and material consumption. All main process stages are included in the tool, and comprehensive system boundaries are applied throughout the model. The process-model core of the mineral carbonation LCA tool includes 8 process stages defined generically to be applicable to a variety of mineral carbonation technologies (see Figure 1). Because our tool aims to compare process schemes that vary significantly in input resource and process design, the model is built at a general, first-order level. The methodology accounts for the following three types of energy consumption: on-site energy consumption, energy of material and energy inputs consumed in the sector of interest (embodied direct energy), and energy of material and energy inputs consumed in all other sectors (embodied indirect energy). Including both on-site and embodied energy allows a full accounting of the total greenhouse gas (GHG) reduction benefits of each process scheme.

**Figure 1.** Life cycle process model schematic for aqueous mineral carbonation based on the Olivine – 155 ûC case; line thickness is scaled to the energy and mass fluxes.

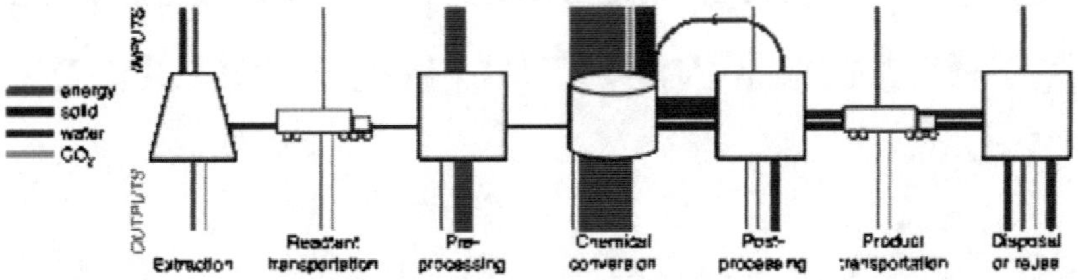


The life-cycle assessment of aqueous mineral carbonation suggests that a variety of alkalinity sources and process configurations are capable of net $CO_2$ reductions. The maximum carbonation efficiency, defined as mass percent of $CO_2$ mitigated per $CO_2$ input, was 83% for CKD at ambient temperature and pressure conditions. In order of decreasing efficiency, the maximum carbonation efficiencies for the other alkalinity sources investigated were: olivine, 66%; SS, 64%; FA, 36%; and serpentine, 13%. Within the range of parameters tested, for all cases maximizing the extent reacted of the alkalinity source unambiguously improves process efficiency. In general, maximizing the extent reacted is critical to optimizing the processes because it minimizes the material handling requirements which contribute negatively to the energy budget. In particular, we observed that:

á the process efficiency is maximized by increasing extent reacted through the most energetically favorable enhancement measure;

á mixing, heating, and grinding are the main energy drivers across all processes;

á reuse of carbonate as aggregate does not necessarily improve life-cycle energy efficiency;

á not all alkalinity sources benefit from high reaction temperatures;

á any steps to increase reaction rates would dramatically improve process efficiency.

Additionally, the $CO_2$ storage potential of mineral carbonation was estimated using the life-cycle assessment results and alkalinity source availability. The annual storage potential for a given alkalinity source was calculated by multiplying its availability (Mt/yr) by the $CO_2$ sequestration efficiency of mineral carbonation of that alkalinity source (t-$CO_2$/t-alkalinity source). For industrial alkalinity sources, availability was based on U.S. production rates (Kelly et al., 2011). For natural alkalinity sources, availability is estimated based on U.S. production rates of a) lime (18 Mt/yr) or b) sand and gravel (760 Mt/yr) (USGS, 2011). The low estimate assumes the maximum sequestration efficiency of the alkalinity source obtained in the current work and the high estimate assumes a sequestration efficiency of 85%. The total $CO_2$ storage potential for the alkalinity sources considered in the U.S. ranges from 1.3% to 23.7% of U.S. $CO_2$ emissions, depending on the assumed availability of natural alkalinity sources and efficiency of the mineral carbonation processes.

### References

Metz, B., O. Davidson, H. C. de Coninck, M. Loos, and L. A. Meyer (eds.) (2005) IPCC Special Report on Carbon Dioxide Capture and Storage, Prepared by Working Group III of the Intergovernmental Panel on Climate Change. Cambridge University Press, Cambridge, United Kingdom and New York, NY, USA.

Kirchofer, A., A. R. Brandt, S. Krevor, V. Priogiobbe and J. Wilcox (2011) Impact of alkalinity sources on the life-cycle energy efficiency of mineral carbonation technologies. In preparation.

Kelly, K. E., G. D. Silcox, A. F. Sarofim and D. W. Pershing (2011) International Journal of Greenhouse Gas Control, **5**, 1587-1595.

USGS (2011) Mineral Commodity Summaries 2011, United States Geological Survey.

# (481d) Uncertainty Quantification in CO2 Sequestration Using Surrogate Models From Polynomial Chaos Expansion

Yan Zhang and Nick Sahinidis, Chemical Engineering, Carnegie Mellon University, Pittsburgh, PA

In this paper, surrogate models are iteratively built using polynomial chaos expansion (PCE) and detailed numerical simulations of a carbon sequestration system. Output variables from the numerical simulator are approximated as polynomial functions of uncertain parameters. Once generated, PCE representations can be used in place of the numerical simulator and often decrease simulation times by several orders of magnitude. However, PCE models are expensive to derive unless the number of terms in the expansion is moderate, which requires a relatively small number of uncertain variables and a low degree of expansion.

To cope with this limitation, instead of using a classical full expansion at each step of an iterative PCE construction method [1], we introduce a mixed-integer programming (MIP) formulation to identify the best subset of basis terms in the expansion. Our approach makes it possible to keep the number of terms small in the expansion. We compare the proposed MIP-based subset selection method against the classical stepwise regression algorithm for building sparse PCE models [2]. The PCE models from our proposed method are simpler than the ones using stepwise regression.

Monte Carlo (MC) simulation is then performed by substituting the values of the uncertain parameters into the closed-form polynomial functions. Based on the results of MC simulation, the uncertainties of injecting CO2 underground are quantified for a saline aquifer. Moreover, based on the PCE model, we formulate an optimization problem to determine the optimal CO2 injection rate so as to maximize the gas saturation (residual trapping) during injection, and thereby minimize the chance of leakage.

References cited:

1. Oladyshkin, S.; Class, H.; Helmig, R.; Nowak, W. An integrative approach to robust design and probabilistic risk assessment for CO2 storage in geological formations. Computational Geosciences 2011, 15, 565–577.

2. Blatman, G.; Sudret, B. An adaptive algorithm to build up sparse polynomial chaos expansions for stochastic finite element analysis. Probabilistic Engineering Mechanics 2010, 25, 183-197.

# (481e) Geo-Chemo-Mechanical Studies for Permanent CO2 Storage in Geologic Reservoirs

**Greeshma Gadikota**[1], Juerg Matter[2], Peter Kelemen[2] and Ah-Hyung Alissa Park[3], (1)Chemical Engineering, Columbia University, New York, NY, (2)Earth and Environmental Sciences, Columbia University, Palisades, NY, (3)Earth and Environmental Engineering & Chemical Engineering, Columbia University, New York, NY

Increasing concentration of $CO_2$ in the atmosphere is attributed to rising consumption of fossil fuels around the world. The development of solutions to reduce $CO_2$ emissions to the atmosphere is one of the most urgent needs of today's society. $CO_2$ injection into geological formations is one of the $CO_2$ storage options with a large capacity. If non-carbonate minerals such as basalt and olivine exist in the $CO_2$ injection site, the injected $CO_2$ can react with those minerals to form thermodynamically stable solid carbonates. In nature, mineral carbonation takes place via dissolution of the mineral to release Ca/Mg ions followed by precipitation of Ca/Mg in the presence of $CO_2$. The rate limiting step of this carbon mineralization process is considered to be the dissolution of the mineral.A question has been raised regarding the effect of in-situ carbon mineralization on the stability of the geologic storage of $CO_2$. Both the kinetics and extents of mineral dissolution and carbonation reactions within the geological reservoir are important factors to evaluate for the long term stability of the geologically injected $CO_2$.Therefore, this work focuses on understanding $CO_2$-mineral-water interactions at a wide range of reaction conditions ($T_{max}$ = 200°C and $P_{CO2, max}$ = 200 bar). Both deionized water and synthetic brine were used as the aqueous reaction media. Minerals such as olivine ($Mg_2SiO_4$), labradorite ($(Ca,Na)(Al,Si)_4O_8$) and basalt (mixture of silicate minerals), were selected based on their abundance in nature and both dissolution and single-step carbonation experiments were performed to investigate fast and long-term kinetics of mineral weathering. Two different regimes (i.e., surface reaction vs. mass transfer limited regimes) of mineral weathering were identified based on the analyses of the activation energy and reaction kinetics.

# (481f) Comparison of CO2 Storage Resource Methodologies

**Angela Goodman**, National Energy Technology Laboratory, Pittsburgh, PA and Grant Bromhal, National Energy Technology Laboratory, Morgantown, WV

Estimating storage resources for geologic $CO_2$ sequestration requires a high-level, first-cut assessment of potential reservoirs that can serve as viable storage sites. $CO_2$ storage resource estimates are based on physically accessible $CO_2$ storage volume in formations without consideration of regulatory or economic constraints. These high-level resource assessments may be used for broad energy-related government policy and business decisions. Subsurface storage volume estimates depend on geologic properties (area, thickness, and porosity of formations) and the efficiency of storage (the fraction of the accessible pore volume that is available to injected $CO_2$). Currently there is no standard method to estimate CO2 storage potential and reported CO2 estimate widely vary. In this work, publically available resource methodologies for predicting how much CO2 can be stored in deep saline formations are compared. Resource methodologies applied are from the United States Department of Energy, United States Geological Survey, Carbon Sequestration Leadership Forum, Massachusetts Institute of Technology, and Lawrence Berkeley National Laboratory. The methodologies are applied to formations throughout the United States.

# (542a) Development of a Novel Hot-Carbonate Process for Post-Combustion CO2 Capture: Role of Organic Promoters in Enhanced Absorption and Bicarbonate Crystallization

**Manoranjan Sahu**, Advanced Energy Technology Initiative, University of Illinois at Urbana-Champaign, Champaign, IL, Qing Ye, University of Illinois at Urbana-Champaign, Champaign, IL and Yongqi Lu, Prairie Research Institute, University of Illinois at Urbana-Champaign, Champaign, IL

Development of cost-effective new technologies are essential for separating and capturing $CO_2$ from power plant flue gases and retrofitting the $CO_2$ capture and sequestration (CCS) technology with minimum cost penalty. Currently available monoethanol amine (MEA)-based processes are highly energy-intensive and expensive. A novel hot-carbonate absorption process with bicarbonate crystallization-enabled high pressure stripping (Hot-CAP) is being developed to reduce the energy requirement for $CO_2$stripping and compression.

In the Hot-CAP, $CO_2$ is absorbed into a concentrated potassium carbonate solution (about 40wt%) at elevated temperatures (60-80°C). A carbonate–based system tends to have a slower $CO_2$ absorption rate compared with the MEA. A key technical issue is thus to enhance the absorption rate to a level that is comparable to MEA-based processes. In this study, five organic catalysts and three potassium amino acid salts were investigated as organic promoter to accelerate the absorption rate of $CO_2$ into the carbonate solution. The rate of $CO_2$ absorption was measured using a stirred tank reactor (STR) reactor. Results indicated that some selected organic catalysts enhanced the absorption rate significantly and the absorption rates were comparable with that of 5M MEA. The absorption rate depended on the promoter dose and $CO_2$ loading level in the solution. Additionally, the crystallization of the $CO_2$ rich solution without and with organic promoters was further investigated under various process conditions. Results from crystallization revealed that potassium bicarbonate prevailed in crystal solids formed.

This presentation will provide a summary of results from the studies of absorption and crystallization in the presence of promoters.

# (542b) Process Development for Phase-Changing Aminosilicone-Based CO2-Capture Absorbents

**Tiffany Westendorf**[1], Benjamin R. Wood[1], Sarah Genovese[1], Irina Sultanova[1], Robert J. Perry[1], Rachel Farnum[1], Michael J. O'Brien[1], Malgorzata Rubinsztajn[1], Ravi-Kumar Vipperla[2] and Robert Enick[3], (1)Global Research, General Electric, Niskayuna, NY, (2)GE Energy, Greenville, SC, (3)University of Pittsburgh, Pittsburgh, PA

Liquid aminosilicone-based absorbents have been developed that have the potential to significantly reduce the cost of $CO_2$ capture from pulverized coal power plants. Upon exposure to $CO_2$, these novel absorbents solidify, thereby enabling innovation in $CO_2$ capture process design. Compared to MEA-based $CO_2$ capture absorbents, the phase-changing aminosilicones have the potential to reduce $CO_2$ capture cost through higher $CO_2$ absorption capacity, greater thermal stability, and reduced solvent losses due to volatility. Furthermore, the phase-changing behavior of these new absorbents and the absence of aqueous diluent enable the use of more energy-efficient separations process units. Finally, by enabling pressurized desorption of $CO_2$, the energy required to compress the clean $CO_2$ stream pre-sequestration is significantly reduced.

A process is being developed in which $CO_2$ is removed from simulated flue gas in a spray absorber, and the solid $CO_2$-sorbent complex is slurried in liquid absorbent for transport to a pressurized desorber. In the desorber, the solids are heated to release $CO_2$ and regenerate the liquid absorbent for recycle. Experiments at the lab scale focusing on spray absorption, slurry rheology, and pressurized desorption are in progress to generate data needed to scale up the process. These results and their effect on estimated $CO_2$ capture economics will be discussed.

# (542c) A Techno-Economic Evaluation of Various Stripper Configurations in Amine Absorption Technology for Post-Combustion Carbon Capture

Sumedh S. Warudkar, Kenneth R. Cox, Michael S. Wong and George J. Hirasaki, Chemical and Biomolecular Engineering, Rice University, Houston, TX

As a part of this work, we are comparing the techno-economic performance of the amine absorption process under a variety of stripper operating conditions – ranging from vacuum strippers operating at 30 kPa to high pressure strippers operating at 300 kPa. We are modeling the amine absorption process using the process simulation software ProMax® with TSWEET®. The equipment sizing and capacity data thus generated are the feed parameters for Aspen Process Economic Analyzer which provides the cost estimates for the process equipment.

In our previous studies, we have reported the comparisons of energy consumption and absorber-stripper column sizing for monoethanolamine (MEA), diethanolamine (DEA) and diglycolamine (DGA) at varying concentrations and stripper operating conditions [1]. Of the three amines compared, we've identified 40 wt% DEA and 60 wt% DGA as the more suitable amines for carbon capture application than MEA. We have also shown that an increase in the stripper operating pressure results in a reduced reboiler energy duty and a smaller stripper column [2]. Use of very low pressure steam - waste heat, results in a reduced 'parasitic energy duty' at the power plant but results in larger stripper columns and increased load on the compression train.

It can be seen that there are two contrasting factors involved in deciding the optimal operational configuration – potential energy savings by reduction of parasitic duty and an increase in the capital expenditure with increasing the equipment sizes. Thus, in spite of the detailed technical analysis conducted previously; it hasn't been possible to predict the optimal configuration amongst those compared. The aim of the techno-economic analysis undertaken as a part of this study is to predict with some clarity the better performing absorbent and the corresponding system configuration.

**References**

1. S. S. Warudkar, K. R. Cox, M. S. Wong and G. J. Hirasaki, "Comparison of alkanolamines for post-combustion carbon capture from a pulverized coal fired power plant", AIChE Fall Meeting, Minneapolis, MN (2011).

2. S. S. Warudkar, K. R. Cox, M. S. Wong and G. J. Hirasaki, "Amine absorption with vacuum stripping for post-combustion carbon capture", AIChE Spring Meeting, Houston, TX (2012).

# (542d) Novel Solvent Selection and Solvent Stripping for CO2 Capture From Power Plants

**Juan M. Salazar**[1], Urmila Diwekar[1], Kevin G. Joback[2] and Abhoyjit Bhown[3], (1)Vishwamitra Research Institute, Center for Uncertain Systems: Tools for Optimization and Management, Clarendon Hills, IL, (2)Molecular Knowledge Systems, Inc., Bedford, NH, (3)Electric Power Research Institute, Palo Alto, CA

**Novel Solvent Selection and Solvent Stripping for $CO_2$ Capture from Power Plants**

Juan Salazar and Urmila Diwekar

Center for Uncertain Systems: Tools for Optimization & Management

Vishwamitra Research Institute

Clarendon Hills, IL 60514

E-mail: urmila@vri-custom.org; Tel: 630-886-3047

And

Kevin Joback

Molecular Knowledge Systems, Inc.

Bedford, NH 03110-0755

And

Abhoyjit S. Bhown

Electric Power Research Institute, Inc.

Palo Alto, CA 94304

This paper presents a new approach to solvent selection for post-combustion $CO_2$ capture from fossil fuel fired power plants. Two main questions arise in a solvent capture process: how to select an effective separating agent and how to design and synthesize this separation process. In previous studies, few solvents were considered for $CO_2$ capture. Further, researchers arrived at these solvents using experimental methods with limited capacity to screen a large number possible candidates. In this work, we are using computer aided molecular design (CAMD) method to obtain a large number of possible solvent candidates. These candidates are then evaluated by integrating the solvent selection and $CO_2$ capture process to ensure improved energy, environmental, and economic performance. In this paper we present the first attempt at using group contribution based CAMD to derive new solvents for $CO_2$ capture.

$CO_2$ absorption in a solvent-based capture process is essentially based on the reversible, selective nature of the chemical reaction between the liquid solvent and $CO_2$ in the flue gas. The process flow diagram shown in Figure 1 has been widely used in literature. Flue gas from the boiler is brought into contact with the solvent in the absorber after the removal of impurities such as NOx, SOx and particulate matter using processes that also cool the flue gas. A blower is used to compensate the pressure drop experienced in the absorber. In the absorber, $CO_2$ selectively absorbs into the solvent, and flue gas leaving the absorber is relatively free of $CO_2$. The $CO_2$-rich solvent from the absorber is then pumped to the top of a stripper (or regeneration vessel), via a heat exchanger. Solvent is regenerated in the stripper at elevated temperatures and near atmospheric pressure. The desorption heat required for removing the absorbed $CO_2$ is provided to the reboiler section of the stripper. The $CO_2$-lean solvent, containing far less $CO_2$ is then pumped back to the absorber via a lean-rich heat exchanger to cool it to the operating temperature of the absorber. We used this process diagram to evaluate new solvents.

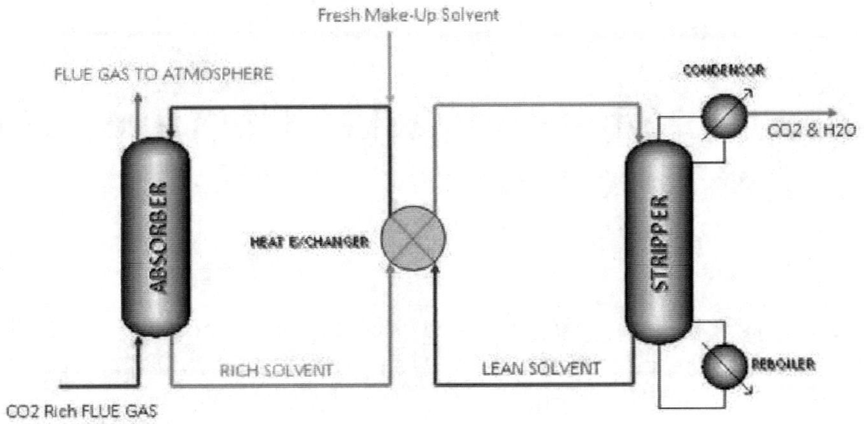


Figure SEQ Figure * ARABIC \s 1 1
Base Absorption Process Flowsheet for $CO_2$ Capture

In our CAMD approach, we concentrated on amines. In order to use group contribution methods, at first we analyzed the equilibrium performance for these new amines for absorption ($CO_2$ capture) and for regeneration (solvent stripping). The CAMD resulted in a list that included a group of more than 50 alkyl alcohol amines whose $CO_2$ solubility properties have not been experimentally determined as far as we know. We evaluated these solvents using ASPEN Plus models and group contribution methods like UNIFAC to evaluate the nonideality. Figure 2 08D0C9EA79F9BACE118C8200AA004BA90B02000000080000000E0000005F005200650066003300320033003000320037003400390030000000 shows the preliminary results of these solvents. The figure shows the energy requirement versus boiling point for 18 solvents simulated both as ideal an nonideal solvents. In general, we observe the nonideality related to these solvents reduces the energy consumption. We will be generating and evaluating additional solvents in order to determine optimum solvents.

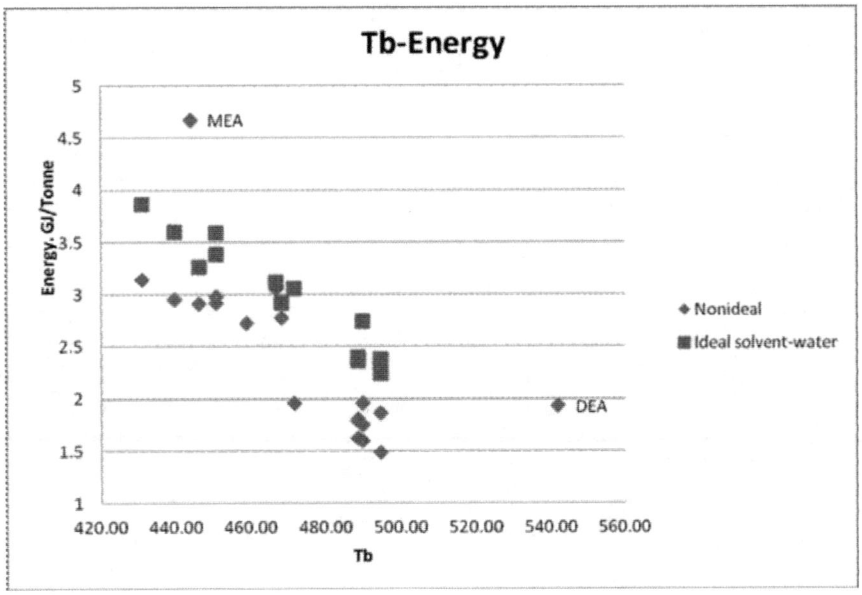


Figure SEQ Figure * ARABIC \s 1 2
Energy Requirements for the New Solvents

# (542e) Low Energy Demanding CO2 Capture Process Using Aqueous Amine Solvent

**Huiyong Kim**, Sogang.ac.kr, Seoul, South Korea and Kwang Soon Lee, Sogang University, Seoul, South Korea

Amine based CO2 capture process has been studied and intensified using the heat integration concept. While many studies on aqueous absorbent-based carbon capture process focus on developing new solvent system, there is much less studies on process intensification to save required energy for CO2 separation.

The purpose of this research has been placed on development of a new process configuration that can achieve a considerable energy saving for CO2 capture. Sensitivity analysis for the absorber and desorber which are two main units of the process provided ideas and possible routes for process modification. With these possibilities, heat flows of the original process were studied for further heat integration to minimize unnecessary heat or coolant usage. With some modifications of the process configuration, heat requirements for solvent regeneration has been reduced considerably. Since minor modifications are proposed, results of this research are belived to be easily applicable to current amine-based carbon capture facilities. The new process has been suggested through a rate-based simulation study using AspenPlus.

# (542f) Designing Reversible Ionic Liquids for CO2 Capture

**Jackson R. Switzer**[1], Amy L. Rohan[1], Kyle M. Flack[2], Emily C. Nixon[2], Amber C. Rumple[2], Elizabeth J. Biddinger[1], Pamela Pollet[2], Charles L. Liotta[2] and Charles A. Eckert[1], (1)School of Chemical and Biomolecular Engineering, Georgia Institute of Technology, Atlanta, GA, (2)School of Chemistry and Biochemistry, Georgia Institute of Technology, Atlanta, GA

Reversible ionic liquids (RevILs) have been developed as alternatives to traditionally cited $CO_2$ capture solvents such as monoethanolamine (MEA). RevILs capture $CO_2$ through chemical reaction of $CO_2$ with an amine functionality to form an ionic liquid. The resulting ionic liquid is then capable of absorbing additional $CO_2$ through physical absorption. Physical absorption is attractive because it has a relatively low energy penalty for reversal. RevILs also present additional energy advantages over MEA. Whereas MEA is used as a dilute aqueous solution, reversible ionic liquids are designed to be non-aqueous solvents, utilizing only the residual water present in the emission source. This strategy allows for significant energy savings by eliminating the need to heat large amounts of unused water during regeneration, improving the overall efficiency of the $CO_2$ capture process.

The design of reversible ionic liquids has been an important focal point in our research group. Using structure-property relationships we have shown an ability to tune the relevant $CO_2$ capture properties by altering the structure of the starting RevIL. Properties targeted include $CO_2$ capture capacity, ionic liquid viscosity, temperature of $CO_2$ release, temperature of RevIL evaporation, and the enthalpy of $CO_2$ absorption. In example, we have observed that by varying the degree and length of alkyl chain branching in the RevIL, we can favorably affect $CO_2$ capacity, ionic liquid viscosity, as well as the temperatures of $CO_2$ release and RevIL evaporation. We have also examined several additional modifications to the structure of our RevILs and investigated their effects on $CO_2$ capture properties. Our progress on developing these structure-property relationships to further the design of optimized solvents for $CO_2$ capture will be discussed.

# (542g) CO2 Post Combustion Capture Using Aminosilicone Solvents

**Irina Sultanova**[1], Benjamin Wood[2], Sarah E. Genovese[1], Paul Wilson[3], Rachel Farnum[1], Paul Buckley[3] and Anthony Furman[3], (1)Global Research, General Electric, Niskayuna, NY, (2)General Electric, Niskayuna, NY, (3)GE Global Research Center, Niskayuna, NY

**$CO_2$ post combustion capture using aminosilicone solvents**

Irina Sultanova, Benjamin R. Wood, Sarah E. Genovese, Paul Wilson, Rachel Farnum, Paul Buckley, Anthony Furman (GE Global Research),

A post-combustion $CO_2$ capture process using aminosilicone solvents offers advantages over conventional MEA-based aqueous solvent methods. The aminosilicone-based process can potentially decrease the overall size of the system, decrease the amount of required energy for desorption, and therefore reduce the overall cost of $CO_2$ capture process. A bench-scale system is currently being developed to demonstrate the technical feasibility of continuous $CO_2$ absorption and desorption with liquid aminosilicones, including recovery and reuse of the solvent. Upon completion of the design and construction, the system will be used to generate the data needed for further scale-up of the process. Some physical properties of the aminosilicones including surface tension, viscosity, heat of $CO_2$ absorption, and desorption and absorption isotherms were measured and are currently used in design of unit operations. These results and their effect on the cost of $CO_2$ capture process will be discussed.

# (556a) Electrochemistry and Transport Limitations of Non-Aqueous Li-Air Batteries From First-Principles

**Venkatasubramanian Viswanathan**[1], Jens Hummelshoj[2], Alan Luntz[3] and Jens K. Norskov[1], (1)Chemical Engineering, Stanford University, Stanford, CA, (2)SLAC National Laboratory, Menlo Park, CA, (3)IBM Almaden, San Jose, CA

Li-air batteries have much higher gravimetric energy storage density compared to all other battery chemistries, and this has led to a strong interest in seeing if such batteries could be developed for powering EVs, enabling driving ranges comparable to gasoline powered automobiles. However, many fundamental challenges exist in these batteries that need to be solved before these batteries can become practical.

In this work, we will discuss a detailed electrochemical model describing the discharge process of the growth of $Li_2O_2$ on terraces, steps and kinks. The different kinds of growth mode along the different kinds of sites have vastly different thermodynamic overpotential. This leads to an explanation of the high Tafel slope observed in Li-air batteries. We will also discuss the effect of deep discharge and the origin of 'sudden death' in these lithium-air batteries and we will show that the origin of this 'sudden death' in flat electrode cells is due to the limitation imposed by electron transport to support the required electrochemistry. Using a simple metal-insulator-metal model, we show that discharge products beyond > 10 nm, have serious transport limitations and this leads to a bias rise across the film. In addition, we also discuss the effects of the presence of a carbonate layer which has been recently identified experimentally. We will also discuss the development of a model incorporating details from some of the recent experiments that show the presence of a carbonate layer. We conclusively show that the origin of potential rise during recharge of the lithium-air batteries is due to a rising concentration of carbonate impurities at the $Li_2O_2$-electrolyte interface. This rise in the overpotential leads ultimately to the decomposition of the electrolyte and limits the rechargability of lithium-air batteries.

# (556b) Nanofiber-Based Novel Electrode Architectures for Batteries and Supercapacitors

Chau Tran, Nataliia Mozhzhukhina and **Vibha Kalra**, Chemical and Biological Engineering, Drexel University, Philadelphia, PA

The ability to control the nano-architecture within electrode materials is critical to the development of high performing energy storage devices. We will present our work on development of nanofiber-based hierarchically structured materials for application in lithium-air batteries and electric double-layer capacitors. For supercapacitors, we report a facile method for obtaining extremely high surface area (>1500 $m^2/g$) and uniformly porous carbon nanofibers. As a first step, blends of polyacrylontritrle (PAN) and a sacrificial polymer in dimethyl formamide (DMF) were electrospun into non-woven nanofiber mats with diameters in the range of 200-300 nm. Fast evaporation of solvent (~200 nl/s) and high elongational flow rate (~$10^5$ $s^{-1}$) during electrospinning allowed us to prevent phase separation and develop a co-continuous morphology of PAN and the sacrificial polymer in the nanofibers. As a second step, electrospun nanofiber mats were subjected to stabilization and carbonization processes to obtain porous carbon nanofibers (CNFs) as PAN converted to carbon and the sacrificial polymer decomposed out to create intra-fiber pores. Unlike other studies so far, we chose the sacrificial polymer possessing a high decomposition temperature and chain rigidity which prevented fibers from shrinking and collapsing during the PAN stabilization process at ~280 °C. Here we demonstrated that using Nafion as a sacrificial polymer allowed us to obtain CNFs with specific surface area of up to 1600 $m^2/g$ without any activation process. We exhibit the tunability of the pore sizes within CNFs by varing material composition. These materials exhibit a large specific capacitance (>200 F/g and >30 $F/cm^3$) possibly due to the presence of a large fraction of meso-pores (2-4 nm) compared to activated carbons (pores <2 nm), which leads to an increase in the accesible carbon surface, thereby improving specific capacitance.

For lithium-air batteries, we incorporate (via in-situ synthesis) manganese dioxide as the catalyst in porous carbon nanofibers to fabricate multi-functional air cathodes. The aim is to establish well-defined multi-phase boundaries within the cathodes such that all reactants; oxygen (via electrode pores), lithium ions (via electrolyte in pores) and electrons (via carbon) can simulatneously reach the catalyst nanoparticles for the oxygen reduction reaction to take place efficiently. Strutural characterization of all nanofiber samples was conducted using scanning electron microscopy (for external nanofiber morphology), transmission electron microscopy of microtomed fiber sections (for internal structure), and nitrogen sorption isotherms (for pore size distribution, pore volume and surface area). Electrochemical performance was studied using cyclic voltammetry at different scan rates, galvanostatic charge-discharge measurements at different current densities, and electrochemical impedance spectroscopy (EIS).

# (556c) Hybrid Li-Air Battery with Sulfuric Acid Electrolyte and Buckypaper Air Cathode

**Yunfeng Li**, Chemical and Biological Engineering, Missouri University of Science and Technology, Rolla, MO, Kan Huang, Chemical and Biological Engineering, Missouri University of Science and Technology, rolla, MO and Yangchuan Xing, Dept. of Chemical and Biological Engineering, Missouri University of Science and Technology, Rolla, MO

Li-air battery (LAB) with aqueous-based catholyte and a water-stable lithium electrode using a LISICON-type lithium ion conducting glass ceramic membrane can eliminate the problems of Li-air battery with organic electrolytes, such as air-breathing cathode choking and degradation of Li anode. We present a rechargeable dual-electrolyte hybrid Li-air battery (HyLAB) with outstanding discharging performances has been demonstrated. The HyLAB has sulfuric acid as the catholyte and has a large active electrode area of 10 $cm^2$. Two types of air-breathing cathode with active layer made from a buckypaper with 5 wt.% Pt supported on carbon nanotubes (Pt/CNTs) and 5 wt.% Pt on carbon black (Pt/CB) as electrocatalyst for oxygen reduction were prepared, and their cell discharge performances were also compared. The cell with Pt/CNTs showed a much better performance than the cell with Pt/CB at high current densities. A better performance of Pt/CNTs was attributed to the more open porous structures formed in the fibrous CNTs. The HyLAB with Pt/CNTs can achieve a discharging capacity of 306 mAh/g at a cut-off voltage of 3.15 V and current density of 0.2 $mA/cm^2$. The corresponding specific energy is 1067 Wh/kg based on the total weight of the sulfuric acid. Detailed results will be presented in this presentation.

# (556d) Modeling Phase Separating Materials in Porous Electrodes Using Non-Equilibrium Thermodynamics

**Todd R. Ferguson**, Chemical Engineering, Massachusetts Institute of Technology, Cambridge, MA and Martin Z. Bazant, Chemical engineering and Mathematics, MIT, Cambridge, MA

Phase separating materials are prevalent in lithium-ion batteries. Commercial lithium-ion battery anodes typically consist of graphite, a material known to go through multiple "stages" as the filling fraction (i.e. percentage) of lithium in the material changes. Other phase separating materials, such as lithium iron phosphate (LFP), have gained a lot of interest due to rate and capacity improvements.

Despite their widespread use, the behavior of phase separating materials inside batteries isn't very well known. The disappearance of voltage plateaus with increased discharge rates and the existence of a non-vanishing voltage gap in the limit of zero current are two phenomena which are not very well understood.

We present a thermodynamically consistent form of porous electrode theory. This set of equations, derived from concentrated solution theory using non-equilibrium thermodynamics and a general reaction rate equation, is used to model phase separating materials in lithium-ion batteries. These models demonstrate phenomena often observed in phase separating battery materials, such as a non-vanishing voltage gap, the mosaic effect, current dependent suppression of phase transformation, and porous electrode effects in phase separating materials (i.e. discrete filling).

This model can be expanded to address a wide range of materials, including LFP and graphite, using thermodynamic models for the materials. The kinetics, equilibrium potential, and particle dynamics are all coupled to the thermodynamics of the material. These dynamics are then coupled to transport in the electrolyte during simulation of the cell, producing a wide variety of behavior typically observed in phase separating materials.

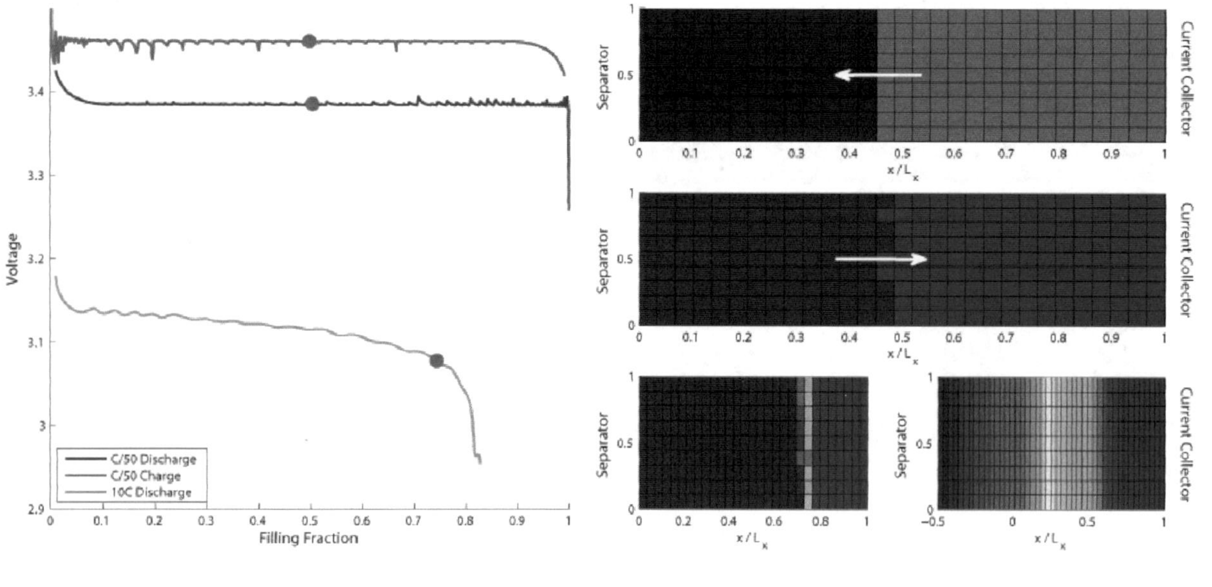

# (556e) Improve the Mass Transfer to Graphite Electrode Through Cell Design in Vanadium Redox Flow Batteries

**Feng Chen**, Pacific Northwest National Laboratory, Richland, WA; Materials Science, Washington State University, Richland, WA

Improve the mass transfer to graphite electrode through cell design in vanadium redox flow batteries

Feng Chen, Yuyan Shao, Qingtao Luo, Wei Wang, Liyu Li, Zimin Nie, Gary Yang

Pacific Northwest National Lab, Richland, WA99354

**Abstract**

Electrolyte mass transfer is a key factor that affects the performance of vanadium redox flow battery (VRFB). Flow batteries with graphite felts inserted with graphite compounded plates were designed to investigate the influence of mass transfer between electrolyte and graphite fiber on the cell performance. Different amounts of graphite plates were applied in the cell design to explore the extent of influence of plate-like barriers on the mass transfer and diffusion rate of electro-active ions. Vanadium redox couples were used as the electro-active species. The flow batteries with new design were also tested under different flow rates to examine the combinational effect of flow rate and graphite plate barriers. Another set of PVC plate barriers were also employed in the new design to study the effect of barrier material on the cell performance. It was shown that the coulombic, energy, and voltage efficiencies were greatly improved by applying the new electrode design in the flow battery system. The testing of cells with PVC insulated plates is still on-going. However, the testing of cell with new design under different flow rates needs to be repeated and verified.

In-Situ Investigation of Vanadium Ions Transport in Redox Flow Battery

*Qingtao Luo,[1] Liyu Li,[2] Zimin Nie,[1] Wei Wang,[1] Xiaoliang Wei,[1] Feng Chen,[1] Bin Li,[1] Guan-guang Xia,[1] Baowei Chen,[1] Zhenguo Yang,[2] Sprenkle Vincent[1]*

[1] Pacific Northwest National Laboratory, P. O. Box 999, Richland, WA 99354, USA

[2] UniEnergy Technologies, LLC, 4333 Harbour Pointe Blvd SW, Unit A, Mukilteo, WA 98275, USA

The all-vanadium redox flow battery (VRB) is probably the most promising and extensively researched flow battery system [1-3]. The ion exchange membrane (IEM) is one of the most important components in a RFB system; it prevents the crossover of the active materials in the positive and negative electrolytes while supporting proton transport to complete the circuit. Perfluorinated polymer membranes (e.g., NAFION_ membranes) have become the choice of IEMs for VRBs due to their high conductivity and excellent chemical stability in highly oxidizing positive electrolytes [4, 5]. However, severe capacity decay of VRBs during cycling were found and reported by several groups [6-8]. Although the capacity fading mechanism is not entirely understood by the VRB research community, the inevitable crossover of vanadium ions across the membrane may play a significant role in causing the capacity fading during cycling. In order to shed light on the VRB capacity fading mechanism, it is necessary to investigate the transport behaviors of vanadium ions across the membrane.

In this work, redox flow batteries with vanadium and iron redox couples were constructed to investigate the transport behaviors of vanadium ions. It was shown that the transport of vanadium ions was affected by a combination of different factors, such as the charge/discharge process, current density, and SOC. The electric field accelerated the positive-to-negative and reduced the negative-to-positive transport of vanadium ions in the charging process and affected the vanadium ion transport in the opposite way during discharge. With the increase of current density, the vanadium ion transport rate increases during the charge process but decreases during discharge. At each given current density, the vanadium ion transport rate, in general, demonstrated a trend of decreasing with the increase of SOC. Furthermore, A simplified Nernster-Planck equation was established to simulate the vanadium ion transport, and the concentration-gradient diffusion coefficients and electric-migration coefficients were calculated on the basis of experimental data.

**Reference**

[1] B. Dunn, H. Kamath, J.-M. Tarascon, Science 334 (2011) 928-935.
[2] Z. Yang, J. Zhang, M.C.W. Kintner-Meyer, X. Lu, D. Choi, J.P. Lemmon, J. Liu, Chem. Rev. 111 (2011) 3577-3613.
[3] M. Skyllas-Kazacos, M.H. Chakrabarti, S.A. Hajimolana, F.S. Mjalli, M. Saleem, J. Electrochem. Soc. 158 (2011) R55-R79.
[4] Q. Luo, H. Zhang, J. Chen, D. You, C. Sun, Y. Zhang, J. Membr. Sci. 325 (2008) 553-558.
[5] X. Li, H. Zhang, Z. Mai, H. Zhang, I. Vankelecom, Energy Environ. Sci. 4 (2011).
[6]S. Kim, J. Yan, B. Schwenzer, J. Zhang, L. Li, J. Liu, Z. Yang, M.A. Hickner, Electrochem. Commun. 12 (2010) 1650-1653.
[7] C. Jia, J. Liu, C. Yan, J. Power Sources 195 (2010) 4380-4383.
[8] T. Sukkar, M. Skyllas-Kazacos, J. Appl. Electrochem. 34 (2004) 137-145

# (556g) Energy Storage in Redox-Active Organic Molecules

**Fikile R. Brushett**, Lu Zhang, John T. Vaughey and Andrew N. Jansen, Chemical Sciences & Engineering Division, Argonne National Laboratory, Argonne, IL

Low-cost, scalable energy storage systems are required to improve the energy efficiency of the electrical grid (*e.g.*, loading-level, frequency regulation) and to enable the widespread penetration of intermittent renewable energy sources (*e.g.*, wind, solar) [1]. Redox flow batteries may provide the best combination of cost, efficiency, and scalability to enable these applications. Present flow battery technologies have low energy densities (< 40 Wh/L), due to the use of aqueous electrolytes with low redox species solubilities (~2 M) and low operating voltages (< 1.5 V, limited by water electrolysis), which, in turn, lead to high system-level costs ($280-450/kWh [2]). Employing non-aqueous electrolytes offers a wider window of electrochemical stability which enables cell operation at dramatically higher potentials (2-4 V) [3-5]. If appropriate redox couples can be identified, operating at higher cell voltages leads to greater system energy (and power) densities and higher roundtrip efficiencies. Moreover, to achieve the same output as an aqueous system, fewer stack layers, lower flow velocities, smaller tanks, and fewer ancillaries would be required, significantly reducing hardware costs and enhancing system reliability.

Redox-active organic molecules may be good candidate materials for non-aqueous flow batteries. This is an attractive approach to energy storage as it enables flexibility to design molecules with favorable properties (*i.e.*, redox potential, activity, solubility) by tailoring either the redox-active moiety or the surrounding molecular structure. Furthermore, research in overcharge protection materials for lithium (Li)-ion batteries has led to the identification of a number of promising materials and to the establishment of some key design principles. Here, we will present our research efforts on a non-aqueous Li-ion redox flow battery employing 2,5-di-tert-butyl-1,4-dialkoxybenzene and quinoxaline as the high- and low-potential active species.

[1] Z. Yang, J. Zhang, M.C.W. Kintner-Meyer, X. Lu, D. Choi, J.P. Lemmon, J. Liu, *Chem. Rev.*, **2011**, *111*, 3577.

[2] D. Rastler, *Electric Perspectives*, **2008**, *Sept/Oct,* 30.

[3] Y. Matsuda, K. Tanaka, M. Okada, Y. Takasu, M. Morita, T. Matsumura-Inoue, *J. Appl. Electrochem.*, **1988**, *18*, 909.

[4] Q. Liu, A.A. Shinkle, Y. Li, C.W. Monroe, L.T. Thompson, A.E.S. Sleightholme, *Electrochem. Commun.,* **2010**, *12*, 1634.

[5] J.Y. Mun, M.-J. Lee, J.-W. Park, D.-J. Oh, D.-Y. Lee, S.-G. Doo, *Electrochem. Solid-State Lett.*, **2012**, *15*, A80.

# (630a) Design, Analysis, Operation and Control of a Photovoltaic/PEM Fuel Cell/Nicd Battery Hybrid Renewable Energy System (HRES) for Urban Transit Applications

**Zachary S. Whiteman**[1], Piyush Bubna[2], Ajay K. Prasad[3] and Babatunde A. Ogunnaike[1], (1)Department of Chemical and Biomolecular Engineering, University of Delaware, Newark, DE, (2)Department of Mechanical Engineering, University of Delaware, Newark, DE, (3)Department of Mechanical Engineering, Univeristy of Delaware, Newark, DE

**Design, Analysis, Operation and Control of a Photovoltaic/PEM Fuel Cell/NiCd Battery Hybrid Renewable Energy System (HRES) for Urban Transit Applications**

Authors: Zachary S. Whiteman, Piyush Bubna, Ajay K. Prasad and Babatunde A. Ogunnaike

A hybrid renewable energy system (HRES), a system consisting of two or more individual renewable energy systems utilized simultaneously to meet an energy demand, potentially allows one to take advantage of complementary strengths of renewable resources in meeting energy demands in both stationary and mobile applications. Because they take advantage of multiple renewable resources, HRESs may represent a cheaper and more dependable alternative to stand-alone renewable energy systems. This presentation is concerned with a hybrid system for mobile applications, specifically, the combination of a roof-installed photovoltaic array with a PEM fuel cell/NiCd Battery bus currently operating shuttle routes on the University of Delaware campus. The primary objectives are: (i) to design the hybrid system (and an appropriate control system) to satisfy required energy demands effectively but with reduced hydrogen consumption; and (ii) to analyze the economic impact of the installation of a photovoltaic array on the PEM fuel cell/NiCd battery bus.

The study is based on actual solar irradiance data collected from a measurement device installed on the roof of the existing HRES bus which was incorporated into a Simulink model developed for the Photovoltaic/PEM fuel cell/NiCd Battery HRES. Control strategies for maintaining the desired state of charge of the NiCd battery while meeting the total energy demand of the bus were designed and implemented on the complete HRES model simulating real shuttle runs under various operating conditions. An economic analysis of the photovoltaic investment required for realizing the HRES process design was also carried out to determine the break-even period for photovoltaic array efficiencies and photovoltaic array surface areas ranging between 5% - 100% and $5m^2$ - $100m^2$ respectively. The presentation will highlight the energy management and operational performance of the control strategies as well as the economic implications of modifying the existing PEM fuel cell/NiCd HRES bus.

# (630b) Issues On Energy Sustainability: Consumption, Efficiency, and Environmental Impact

Leisha Vance, Tarsha Eason and **Heriberto Cabezas**, Office of Research and Development, U.S. Environmental Protection Agency, Cincinnati, OH

One of the more important challenges in achieving sustainability is how to meet the energy consumption needs of a growing human population in the face of increasing economic prosperity. In this talk, we present a very simple but rigorous procedure for exploring the relationship between energy consumption, environmental impact, human population, gross domestic product, and energy efficiency. The underlying theory basically assumes that energy use increases with time in proportion to two factors: gross domestic product and human population. Other assumptions are that energy use has a non-zero impact on the environment and decreases only with energy efficiency. This approach allows us to pose and evaluate: (1) the limitations given the known behavior of societies and economies, practical technology, and the Laws of Thermodynamics and (2) possible pathways to sustainability. The goal is essentially to look for a way to meet expected energy consumption requirements without increasing the environmental impact beyond that seen in an appropriate reference year. We have chosen the year 1977 as our reference year because it is the year when the combined ecological footprint of the human population world-wide was equal to the biocapacity of the Earth. That is, the capacity of the Earth was able to sustain the global demand of human beings as measured by the ecological footprint. Although we use this metric as a benchmark, we are not suggesting that the ecological footprint is necessarily an ideal measure of environmental impact. Rather, we use it simply as a convenient means of establishing a reasonable reference point. Based on the aforementioned assumptions, basic calculus, and published values for energy use, gross domestic product, human population, and energy efficiency over time are used to explore energy scenarios. Because many of the sustainability issues associated with energy use, such as climate change, are global in nature, we begin with an analysis at the global scale. Given the expected growth in World gross domestic product as well as human population growth and energy use over the next 20+ years, our results indicate that the kinds of energy-use efficiencies needed to keep the environmental impact at or below the 1977 level seem unrealistic. From our calculations, energy-use efficiency would need to rise to approximately 78% by 2030 to maintain the 1977 impact level. Not only is this level nearly impossible to achieve technologically and thermodynamically, but the amount of energy consumed per unit of gross domestic product (and per capita) would need to cumulatively decrease by 1.5% to 2% per year beginning in 2012 for environmental impact of energy use not to exceed that of 1977. Again, this will not not an easy task. The situation, although quite grim, is not hopeless, and the feasibility of an integrated approach to managing the environmental impact of energy use using all of the aforementioned options will be further discussed.

# (630c) Surrogate Model Based Optimal Synthesis of Solid Sorbent Carbon Capture Process

**Zhihong Yuan**[1], Alison Cozad[1], Nick Sahinidis[1] and David C. Miller[2], (1)Chemical Engineering, Carnegie Mellon University, Pittsburgh, PA, (2)U.S. Department Of Energy, National Energy Technology Laboratory, Morgantown, WV

Concerns about climate change have led countries to set ambitious targets for the reduction of anthropogenic greenhouse gas emissions and a number of initiatives have been set up to reduce CO2 emissions[1]. Carbon capture, utilization, and storage technology is an essential route to achieving a meaningful reduction of global CO2 emissions in the context of continued fossil fuel use in the power sector. For example, the U.S. Department of Energy (DOE) has initiated the Carbon Capture Simulation Initiative (CCSI) to accelerate the development of carbon capture technology. Solid sorbent based technologies are being used to demonstrate the capabilities of the computational tools being developed under this initiative.

Obviously, process synthesis, which has had a conspicuous impact on the development, design and operation of (petro) chemical processes[2], will play a vital role for the development of a cost-effective carbon capture process. In this study, a superstructure optimization formulation for the optimal synthesis of a carbon capture system has been established, various reactors including fluidized bed reactors and moving bed reactors together with different potential topologies have been included in the optimization formulation. Since detailed first principle models of the reactors are computationally intractable for large scale superstructure optimization, we have developed the Automated Learning of Algebraic Models for Optimization (ALAMO) approach to generate a set of low complexity algebraic surrogate models of different reactors to reduce the computational burden associated with reactor models developed in commercial process simulation packages.

Through solving the above formulated mixed integer nonlinear optimization problem which aims at minimizing the increase in cost of electricity due to capture while achieving 90% capture target the potential equipment configurations, interconnections and design/operation levels of the selected equipment can be identified. This presentation will demonstrate the application of this superstructure-based process synthesis approach for a solid sorbent-based carbon capture system.

**References**

1. MacDowell N, Adjiman CS, Shah N. An overview of CO2 capture technologies. *Energy Environ Sci*. 2010; 3; 1645-1669.
2. Westerberg AW. A retrospective on design and process synthesis. *Comput Chem Eng. 2004; 28: 447-458.*

# (630d) Modeling of Process Steam Consumption in Chemical Batch Plants Based On Standard Production Documentation

Cecilia Pereira[1], Stavros Papadokonstantakis[2] and Konrad Hungerbühler[2], (1)Chemistry and Applied Biosciences, Institute for Chemical- and Bioengineering, Swiss Federal Institute of Technology, Zurich (ETHZ), Zürich, Switzerland, (2)Chemistry and Applied Biosciences, Institute for Chemical- and Bioengineering, Swiss Federal Institute of Technology, Zurich (ETHZ), Zurich, Switzerland

Minimization of energy consumption in the chemical industry is one of the key principles of green chemistry. Although traditionally most efforts on energy saving studies have been done for continuous production processes, batch processes have recently become the focus of many research projects [1-3]. However, a shortcut modeling approach for screening of energy consumption in multipurpose batch plants, based on a simple and general data extraction procedure from process documentation is missing. In this work a new methodology for modeling of steam consumption in multipurpose chemical batch plants based on standard process documentation, rules of thumb, expert opinion, classical thermodynamics, and some model parameterization according to Bieler et al [3] is introduced. Two case studies using production data from multipurpose batch plants of two different chemical companies were carried out for the development and validation of this new shortcut modeling approach. Several reaction and work-up processes performed in different equipments were considered for the modeling of steam consumption, and the model results were validated against reference data in each case study. Additionally, the validation results were used to derive uncertainty intervals for the model outputs based on fuzzy set theory. It has been shown that the documentation based models proposed in this work provide acceptable estimations of steam consumption in multipurpose batch plants, and that the batch-to-batch variability of the steam consumption is well captured by the uncertainty intervals resulting from the validation procedure. The possibility of allocating steam consumption to individual processes or products in multipurpose batch plants by means of this new short cut modeling approach enables the identification of energy optimization potentials.

1. Andrej Szïjjarto, S.P., * Ulrich Fischer, and Konrad Hungerbühler, *Bottom-up Modeling of the Steam Consumption in Multipurpose Chemical Batch Plants Focusing on Identification of the Optimization Potential.* Industrial & Engineering Chemistry Research, 2008. **47**: p. 7323-7334.

2. Bieler, P.S., U. Fischer, and K. Hungerbuhler, *Modeling the energy consumption of chemical batch plants - Top-down approach.* Industrial & Engineering Chemistry Research, 2003. **42**(24): p. 6135-6144.

3. Bieler, P.S., U. Fischer, and K. Hungerbuhler, *Modeling the energy consumption of chemical batch plants: Bottom-up approach.* Industrial & Engineering Chemistry Research, 2004. **43**(24): p. 7785-7795.

# (630e) A Shortcut Graphical and Spreadsheet Technique for Identifying Heat Exchange Networks

**Russell F. Dunn**, Chemical and Biomolecular Engineering , Vanderbilt University, Nashville, TN and Kenneth A. Debelak, Chemical and Biomolecular Engineering, Vanderbilt University, Nashville, TN

Incorporating heat integration within chemical processes continues to be a critical component for future sustainability of large chemical and polymer manufacturing companies. Numerous design methodologies have been developed for the synthesis of heat exchange networks (HENs) over the past three decades. Several of the methods previously developed are based on a two stage targeting approach. The first stage uses the hot and cold composite curves to identify the maximum heat integration possible for a selected group of streams to be cooled and streams to be heated. This approach is incorporated in several commercially available software packages such as Aspen Energy Analyzer. The second stage identifies the network of heat exchangers that meets the target identified in stage one. Several strategies have been employed for identifying the HEN in stage two including, but not limited to, heuristics, mathematical optimization, and superstructure approaches.

This paper will present a new graphical/spreadsheet technique and design strategy for identifying the HEN in stage two for large industrial problems. Advantages of this strategy will be provided along with several case studies that illustrate the effectiveness of this approach. Finally, experiences using this strategy in teaching undergraduate chemical engineering design courses will also be discussed.

# Sustainabilty Assessment and Design of Centralized and Decentralized Energy Solutions

Healthy economic growth in developing countries combined with increasing population and affluence are resulting in significant increase in total energy consumption. At the same time, growing concern about environmental problems such as air pollution, global warming and depletion of fresh water resources is encouraging efforts for supplying this increasing energy demand in an environmentally friendly and sustainable manner. Depletion and price fluctuation of fossil fuels also mandates finding alternative technologies to satisfy this energy demand.

Especially in rural areas of developing countries, the burden of energy poverty reduces opportunities for enhancing their quality of life. This poverty is due to factors such as absence of electrification or inconsistent availability. Utilization of inefficient energy sources like cow dung, kerosene or biomass residues causes local environmental problems such as indoor smoke further triggering health problems and related healthcare costs. Consequently, many efforts are focused on introducing decentralized and cleaner energy solutions. Such solutions are also being adopted by the developed world due to the desire for enhanced sustainability.

Given these reasons, it is essential to design energy systems consisting of centralized and localized options which create the optimum energy mix to meet energy demand in a manner that is environmentally, socially and economically feasible.

In this study, we evaluate the environmental and energetic performances of different energy technologies utilizing thermodynamic methods such as ecological cumulative exergy consumption or emergy analysis and value chain scale life cycle assessment (LCA). These technologies are two centralized coal gasification technologies (conventional and calcium looping technologies) implemented with $CO_2$ capture, decentralized multi-crystalline solar cells, biomass gasification, and floating drum biogas power plants. Furthermore, using a multi-objective mixed integer programming (MIP) framework and setting environmental, economic, social and thermodynamic objectives, we plan to find optimum energy mix for the selected society. This framework is applied to Rampura, a village in Central India, which is representative of a developing society.

Based on our preliminary results based on data from Rampura, biogas power plant is most attractive from a life cycle thermodynamic point of view, followed by solar power plant, biomass gasifier and with least environmental performance of coal technologies.

For rural biogas plant, we calculate its renewability index to be around 75 % for only biogas production and slightly lower for electricity generation from biogas because of machinery and maintenance requirements. A quality corrected thermodynamic return on investment is calculated to be around 5 for both biogas and electricity. A sensitivity analysis identifies manure and brick (for construction) to be the most important resources. The photovoltaic solar power plant is found to be much less attractive than the biogas plant, with the most dominant inputs being resources for open ground mounting, inverter, maintenance inputs. The renewability index of solar panel is around 2.5 % and thermodynamic return on investment is slightly over 1. Emergy analysis results of biomass gasifier unit are similar to that of solar PV, just renewability index being slightly higher. Biomass, transportation of biomass to the plant site and preparation of biomass as feedstock are the dominant inputs to the system. Not surprisingly, coal gasification is least attractive with least percentage of renewability and similar energetic performance to biomass and solar systems from an environmental point of view, but may be essential for satisfying energy needs. For coal gasification processes; mining of coal, or coal input constitutes more than half of the emergy input to the systems.

LCA enables us to capture downstream effects of processes investigated which emergy analysis cannot evaluate. According to our results, solar PV utilizes least amount of land followed by centralized coal gasification technologies and biogas plant. Likewise biogas plant is most water utilizing system followed by solar PV and coal gasification systems. While conventional coal gasification technology has higher level of GWP compared to solar PV and biogas digester technologies, CLP has lowest GWP. This situation points out that improved coal gasification technologies implemented with $CO_2$ capture can significantly ease environmental burden of generating electricity from coal.

Life cycle assessment of biomass gasifier plant is the next steps in this work followed by the formulation of the optimization problem to identify the "optimum" energy mix while accounting for multiple objectives. This framework will be relevant to choosing combinations of centralized and decentralized options in developing and developed economies.

## "Challenges and Opportunities for Enhancing Exergy Efficiency of Industrial Sectors in Developing Economies – Insight From India's Growth"

Sachin B. Jadhao1, Aniruddha B. Pandit1 and Bhavik R. Bakshi2,

1. Institute of Chemical Technology, Mumbai, India

2. William G. Lowrie Department of Chemical and Biomolecular Engineering The Ohio State University, Columbus, OH.

**Abstract**

Given that the development of emerging economies is invariably accompanied by an increase in energy and resource consumption, goals of sustainable development require a parallel increase in resource efficiency. This talk will focus on the trends and opportunities for enhancing the energy and material efficiency of industrial sectors in India. It uses the approach of exergy analysis since it has been found to be an appropriate measure for resource accounting. Studies on industrial sectors have been carried out for several economies and the resulting insight can be used for identifying opportunities for enhancing resource use and reducing environmental impact. The aim of this work is to understand the metabolism of Industry sector of India as it has evolved from a controlled to a more open free market economy, and its implications for sustainable development. We analyzed various industries in India on the basis of, resource consumption and environmental pollution. We estimate second law efficiencies for various industries in India, including Chemical, Iron and steel, Fertilizer, cement, sugar and others. We also estimate the efficiency of the India's industry sector as a whole. Trends of resource consumption and its conversion into useful work are analyzed using exergy analysis and the results are compared with those of other developing and developed economies. In addition, using techniques of network analysis, opportunities for enhancing exergy efficiency are identified.

## Introduction

Industry sector in India is one of the major resource consuming sector. Figure 1 and figure 2 shows the fuel exergy and electric exergy consumed in Industry sector of India since 1970. The physical resources (fuel and non fuel minerals ) consumption in Industry sector in India has been observed to increasing rapidly from the period of start if industrialization and economic liberalization in India. Among the various industries Inron and steel, fertilizer, Chemical and Cement are the major resource consuming industries. In general the industries are evaluated on the basis of First Law efficiency (calculated by comparing the energy content of the useful output with the energy content of the inputs, including materials ), however the first law does not take in to account the quality which leds to the misleading conclusions [1].

Exergy efficiency is the ratio of potentially useful (exergy) output to potentially useful (exergy) input. [1] Second-law efficiency is usually less than first law efficiency.[1] Exergy efficiency captures the degree to which energy inputs are converted into useful work [3]. Exergy analysis provides insights not available via other methods, especially for identifying unit operations with the greatest potential for improvement [2].

A number of efforts have been undertaken to apply exergy analysis to various sectrors in the economy that includes the industrial sector. Since Industry sector accounts for major share in total resource consumption of many economies (particularly with respect to developing economies like India and China), it is important to examine the efficiency of resource utilization in this sector. Many studies in past have analysed the Industrial sector of various economies on the basis of exergy efficiency. These studies presented exergy accounts for individual process stages, in order to identify major losses and evaluate the potential for further technical improvements in the processes. Industrial sector in India have been compared and analyzed on the basis of efficiency or resource conversion in different industries. Most of the existing research on thermodynamic analysis of industry mainly accounts for fossil fuel consumption and its conversion to useful work. Few studies have been conducted using exergy assessment of industries that takes in to account fuel, non-fuel minerals resources. In this study we have analyzed the industries on the basis of fuel and non-fuel resource consumption and emissions.

**Research method**

In this work, conversion and utilization of all natural resources including energy and materials are described in terms of exergy flow in various Industries in India. Exergy utilization trends and efficiencies for various industries are analyzed and compared with those of other studies. Exergy losses in transformation of material resources and in conversion of various forms of energy into useful work are described in detail.

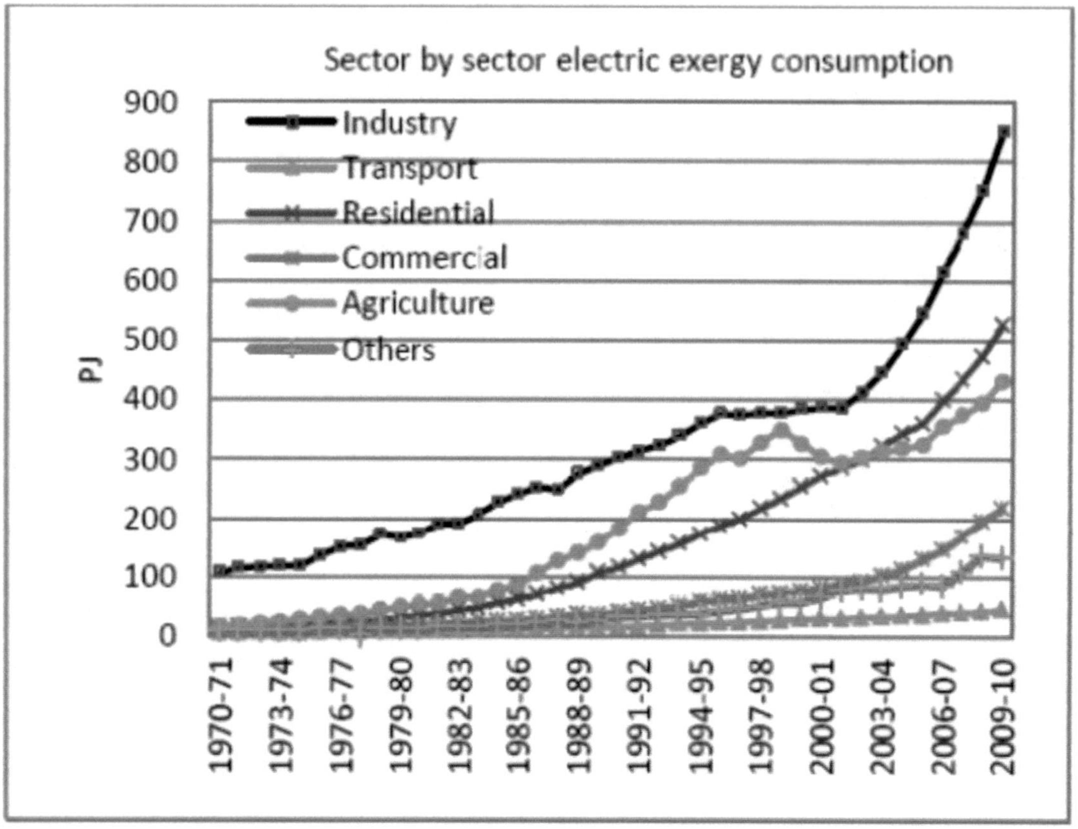


Figure 1 Electric exergy consumed in Industry sector of India

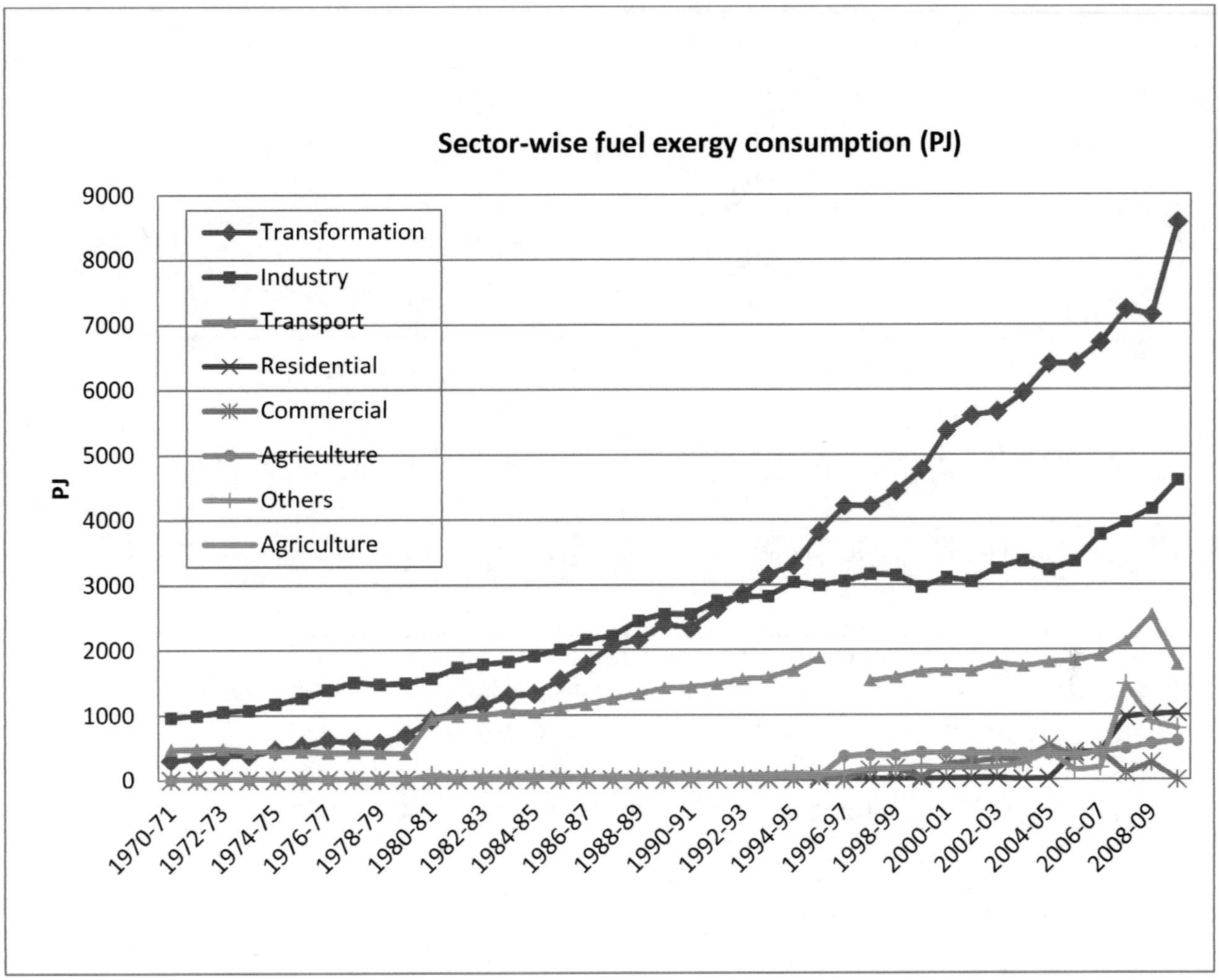


Figure 1 Fuel exergy consumed in Industry sector of India

## Sample calculation of exergy efficiency for Iron & Steel Industry in India (2007-08)

Energy utilization in the industry sector is categorized according to the energy use as follows

| Energy use in the industrial sector in India by type of Use (%) | | |
|---|---|---|
| Use Category | Electrical | All Fuels |
| Mechanical Drive | 65 | 0 |
| Heating | 28 | 100 |
| Others (Assumed Lighting) | 7 | 0 |
| TOTAL | 100 | 100 |

**INPUTS**

Total Electric energy consumed in the Iron & Steel Industry= 75.78 **PJ**

Total Electric exergy consumed in the Iron & Steel Industry= 75.78 **PJ**

Total Fuel energy consumed in the Iron & Steel Industry= 369.21 **PJ**

Total Fuel exergy consumed in the Iron & Steel Industry= 391.37 **PJ**

**[I] Mechanical Drive**

Since 65 % of the electric energy is used for mechanical drive application

$$Energy\ \&\ exergy\ input\ for\ machenical\ drive = 75.78 \times 0.65 = 49.26\ PJ$$

Mechanical drives are assumed to be 90 % efficient

$$Enegy\ \&\ exergy\ output\ for\ machenical\ drive = 49.26 \times 0.90 = 44.33\ PJ$$

**[II] Heating**

Process heating in the industry is subdivided in to three categories based on the temperature ranges as follows. The temperature range and the efficiency values are taken from the study of Dincer et al.

| Process heating temperatures and efficiencies for the industrial sector | | | |
|---|---|---|---|
| Category | Temperature | Electrical | Fuel |

| | (° C) | (%) | (%) |
|---|---|---|---|
| Low | <121 | 100 | 65.5 |
| Medium | 121–399 | 90 | 60 |
| High | >399 | 70 | 50 |

**Breakdown of the energy used for each temperature range in the industry is referred from the Brown et al. study for 108 industrial processes**

From this study the breakdown for Iron & steel Industry is given as follows

| Breakdown of energy used for each temperature (%) | | | | | |
|---|---|---|---|---|---|
| Industry | T range | TO | Mean T (°C) | Electricity | Fuel |
| Iron and steel | Low | 25 | 45 | 4.2 | 0 |
| | High | 25 | 983 | 95.8 | 100 |

Exergy efficiency of low temperature electric heating = 45+273 = 318°K and **η= 100 %**

$$\Psi_{e\,h\,Low} = 1 \times \left[1 - \left(\frac{298}{318}\right)\right] = 6.28\,\%$$

Exergy efficiency of high temperature electric heating: T = 983+273 = 1256 °K and **η= 70 %**

$$\Psi_{e\,h\,high} = 0.7 \times \left[1 - \left(\frac{298}{1256}\right)\right] = 53.39\%$$

Exergy efficiency of low temperature fuel heating: T = 45+273 = 318 °K and **η = 65 %**

$$\Psi_{f\,h\,Low} = 0.65 \times \left[1 - \left(\frac{298}{318}\right)\right] = 4.08\ \%$$

Exergy efficiency of high temperature fuel heating: T = 983+273 = 1256 °K and **η = 50 %**

$$\Psi_{f\,h\,high} = 0.5 \times \left[1 - \left(\frac{298}{1256}\right)\right] = 38.13\ \%$$

Energy efficiency of electric heating= efficiency×percentage of electric energy used

$$\eta_e = 100\% \times 4.2\% + 90\% \times 0\% + 70\% \times 95.8\% = 71.26\%$$

Energy efficiency of fuel heating= efficiency× percentage of fuel energy used in iron & steel industry

$\eta_f$ = 65%× 0%+ 60%×0%+50%×100% = 50%

Exergy efficiency of electric heating= efficiency×% electric energy used in iron & steel industry

$$\Psi_e = 6.28\% \times 4.2\% + 53.39\% \times 95.8\% = 51.0\%$$

Exergy efficiency of fuel heating= efficiency×% fuel energy used in iron & steel industry

$$\Psi_f = 7.2\% \times 0\% + 38.7\% \times 100\% = 38.7\%$$

**In Industry 28 % of the electricity is used for heating applications.**

Energy Efficiency of heating

$$\eta_q = \frac{(Electric\ energy\ used \times efficiency + Fuel\ energy\ used\ \times efficienfy)}{(Electric\ energy\ + Fuel\ Energy\ input)}$$

$$\eta_q = \frac{(75.78 \times 0.28 \times 0.713 + 369.21\ \times 0.50)}{(75.78 \times 0.28\ + 369.21)}$$

$$\eta_q = 51.15\ \%$$

Exergy Efficiency of heating

$$\Psi_q = \frac{(Electric\ exergy\ input \times efficiency + Fuel\ exergy\ input\ \times efficienfy)}{(Electric\ exergy\ + Fuel\ Exergy\ input)}$$

$$\Psi_q = \frac{(75.78 \times 0.28 \times 0.514 + 391.37\ \times 0.387)}{(75.78 \times 0.28\ + 391.37)}$$

$$\Psi_q = 39.35\ \%$$

$$Energy\ Output\ of\ heating = Energy\ Input \times Energy\ Efficiency\ for\ heating$$

$$Energy\ Output\ of\ heating = 75.78 \times 0.28 \times 0.72 + 369.21 \times 0.50 = 199.88\ PJ$$

$$Exergy\ Output\ of\ heating = Exergy\ Input \times Exergy\ Efficiency\ for\ heating$$

$$Exergy\ Output\ of\ heating = 75.78 \times 0.28 \times 0.52 + 391.37 \times 0.387 = 162.35\ PJ$$

**[III] Lighting**

7 % of electricity is used for lighting application

$$Energy\ \&\ exergy\ input\ for\ lighting = 75.78 \times 0.07 = 5.30\ PJ$$

Lighting energy and exergy efficiencies are assumed as **18.5% and 17.1%**

$$Enegy\ output\ for\ lighting = 5.30 \times 0.185 = 0.98\ PJ$$

$$Exegy\ output\ for\ lighting = 5.30 \times 0.17 = 0.90\ PJ$$

$$Total\ Energy\ Input$$
$$= Aggrigate\ energy\ input\ for\ heating, lightening\ and\ mechanical\ drive$$

$$Total\ Energy\ Input = 369.21 + 75.78 = 444.99\ PJ$$

$$Total\ Exergy\ Input = 391.37 + 75.78 = 466.78\ PJ$$

$$Total\ Energy\ output = 199.88 + 44.33 + 0.98 = 245.19\ PJ$$

$$Total\ exergy\ output = 162.35 + 44.33 + 0.90 = 207.58\ PJ$$

$$\boldsymbol{Overall\ energy\ efficiency} = \frac{Total\ energy\ output}{Total\ energy\ input} \times 100$$

$$= \frac{245.19}{444.99} \times 100 = \mathbf{55.10}\ \%$$

$$\boldsymbol{Overall\ exergy\ efficiency} = \frac{Total\ exergy\ output}{Total\ exergy\ input} \times 100$$

$$= \frac{207.58}{466.78} \times 100 = \mathbf{44.34}\ \%$$

**Results and discussion**

This work provides insight into areas where major improvements are required to enhance resource efficiency in the various industries. Effects of factors such as increasing consumption and depletion of resource quality on overall efficiency are demonstrated.

REFERENCES

[1] R. U. Ayres, L. T. Peir, and G. V. Mendez. Exergy e-fficiency in industry where do we stand? Environ. Sci. Technol, 2011.

[2] B. R. Bakshi and G. F. Grubb. Appreciating the role of thermodynamics in lca improvement analysis via an application to titanium dioxide nanoparticles. *Environmental Science and Technology,* 45, 7, 3054–3061, 2011.

[3] E. Williams, B. Warr, and R. U. Ayres. Efficiency dilution: Long-term exergy conversion trends in japan. Environ. Sci. Technol., 42:49644970, 2008.

# (774a) Development of Proton Selective Membranes for Efficient and Economical Vanadium Redox Flow Batteries

**Justin Walls**[1], Pedram Jahanian[2], Guangzhao Mao[2], Dr. Steven Salley[3] and Dr.Simon Ng[3], (1)Chemical Engineering, Wayne State University, Detroit, MI, (2)Chemical Engineering and Materials Science, Wayne State University, Detroit, MI, (3)Chemical Engg., Wayne State University, Detroit, MI

**Development of Proton Selective Membranes for Efficient and Economical Vanadium Redox Flow Batteries**

Justin Walls, Padram Jahanian, Guangzhao Mao, Steven Salley, K. Y. Simon Ng

Department of chemical Engineering and Materials Science, College of Engineering,

Wayne State University, Detroit MI 48202

Flow Redox Batteries (FRBs), especially the Vanadium Redox Flow Battery (VRFB), provide a promising chemistry for storing intermittent renewable energy on a very large scale. While these batteries have already been developed on the megawatt hour level; none of the FRB technologies has seen significant market penetration. This is mainly because the technology is still very expensive, due to high material costs, and less than satisfactory performance parameters and reliability.

The key component of the VRFB, which is one of the leading causes of the elevated cost and unreliability, is the ion exchange membrane (IEM). The membranes need to possess good ion conductivity while maintaining high mechanical and chemical stability. Perfluorosulfonic polymers such as Nafion are the most common membranes for proton exchange in the VRFB due to their good chemical stability and high proton conductivity. However, due to the tendency of vanadium to pass through the membrane, Nafion membranes suffer from decreased energy efficiency. This project is focused on decreasing the vanadium ion diffusion through the membrane by modifying the nanostructure of Nafion. There has been extensive research activity in the modification of Nafion membranes, for example, modification by layer-by-layer polyelectrolyte multilayer films and in situ sol-gel reaction to close the pores by silica nano particles. Our project combines these two methods using hyper-branched polyelectrolyte multilayer films to further lower the vanadium ion diffusion rate. Nafion modification by the sol-gel chemistry was conducted by treating the Nafion in the tetraethylorthosilicate (TEOS) methanol solution. Poly(diallyldimethylammonium chloride) (PDDA) and poly(sodium styrene sulfonate) was used in the layer-by-layer deposition. Diffusivity and ion crossover of these membranes was tested and compared to that of the unmodified membranes as well as other modification techniques that have been reported. A small scale VRFB system was constructed to test the performance of these modified membranes, and efficiency and reliability of the system as a whole was measured.

# (774b) Microporous Separators On Fe/V Redox Flow Battery: A Valuable Opportunity for Cost Reduction

**Xiaoliang Wei**[1], Qingtao Luo[1], Bin Li[1], Zimin Nie[1], Yuyan Shao[1], Feng Chen[1], Baowei Chen[1], Guanguang Xia[1], Liyu Li[2], Zhenguo Yang[2] and Wei Wang[1], (1)Pacific Northwest National Laboratory, Richland, WA, (2)UniEnergy Technologies, LLC, Mukilteo, WA

**Microporous Separators on Fe/V Redox Flow Battery: A Valuable Opportunity for Cost Reduction**

Wei, Xiaoliang; Luo, Qingtao; Li, Bin; Nie, Zimin; Shao, Yuyan; Chen, Feng; Chen, Baowei; Xia, Gordon; Li, Liyu; Yang, Z Gary; Wang, Wei

Pacific Northwest National Laboratory, 902 Battelle Boulevard, PO Box 999, Richland, WA 99354

Redox flow batteries are considered as one of the most promising medium-to-large scale energy storage technologies and have attracted much attention both academically and industrially.[1],[2] A number of redox chemistries have been proposed, and significant progress has been achieved by today.[3],[4] However, broad market penetration of redox flow batteries is still hindered by their intrinsic limitations such as high cost and low durability of components, narrow operational temperature range, limited solubility of active species, and/or low electrochemical activity.[5] The recently invented iron-vanadium (Fe/V) redox flow battery employing $Fe^{2+/3+}$ – $V^{2+/3+}$ redox couples have shown to be a promising option for stationary energy storage.[6],[7] The Fe/V flow battery demonstrated stable cycling performance with a nearly 100% utilization ratio over a broad temperature range of 0-50°C.

The positive charged species, $Fe^{3+}$, is a relatively weak oxidant. Therefore, hydrocarbon-based ion exchange membranes and/or separators are possible options for use in Fe/V flow battery system. This contribution investigated the cycling performance of a variety of polyethylene microporous separators on Fe/V flow cell. Among them, some separators exhibited energy efficiency of around 70% at temperatures ranging from 5-50°C and at current densities up to 80mA/cm². Because these separators are very inexpensive, their use significantly reduces the capital cost of Fe/V flow battery, delivering great potential for developing a low-cost energy storage system.

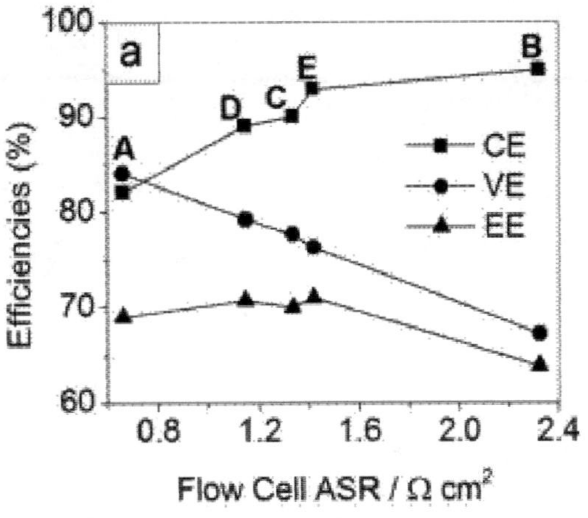


Figure 1 Flow cell cycling efficiencies (CE, VE, and EE) of microporous separators A, B, C, D, and E.

[1] Z. Yang, J. Zhang, M. C. W. Kintner-Meyer, X. Lu, D. Choi, J. P. Lemmon, and J. Liu, *Chem. Rev.* 2011, *111*, 3577-3613.

[2] J. Rugolo, and M. J. Aziz *Energy Environ. Sci.* 2012, **DOI:** 10.1039/C2EE02542F.

[3] M. Skyllas-Kazacos, M. H. Chakrabarti, S. A. Hajimolana, F. S. Mjalli, and M. Saleem *J. Electrochem. Soc.* 2011, *158*, R55-R79.

[4] L. Li, S. Kim, W. Wang, M. Vijayakumar, Z. Nie, B. Chen, J. Zhang, G. Xia, J. Hu, G. Graff, J. Liu, and Z. Yang *Adv. Energy Mater.* 2011, *1*, 394–400.

[5] S. Eckroad, *Vanadium Redox Flow Batteries: An In-Depth Analysis.* EPRI, Palo Alto, CA: 2007. 1014836.

[6] W. Wang, S. Kim, B. Chen, Z. Nie, J. Zhang, G. Xia, L. Li, and Z. Yang *Energy Environ. Sci.* 2011, *4*, 4068-4073.

[7] W. Wang, Z. Nie, B. Chen, F. Chen, Q. Luo, X. Wei, G. Xia, M. Skyllas-Kazacos, L. Li, and Z. Yang *Adv. Energy Mater*. 2012, *2*, 487–493.

# (774c) In Situ Investigation of Vanadium Ions Transport in Redox Flow Battery

**Qingtao Luo**[1], Liyu Li[2], Zimin Nie[3], Wei Wang[1], Xiaoliang Wei[1], Feng Chen[1], Bin Li[4], Guanguang Xia[1], Baowei Chen[5], Gary Yang[1] and Sprenkle Vincent[1], (1)Pacific Northwest National Laboratory, Richland, WA, (2)Hydrocarbon Processing Group, Pacific Northwest National Laboratory, Richland, WA, (3)Pacific Northwest National Labratory, Richland, WA, (4)Pacific Northwest National Laboratory, Richland , WA, (5)Biological Sciences, Pacific Northwest National Laboratory, Richland, WA

The all-vanadium redox flow battery (VRB) is probably the most promising and extensively researched flow battery system. In order to shed light on the VRB capacity fading mechanism, it is necessary to investigate the transport behaviors of vanadium ions across the membrane. In this work, flow batteries with vanadium and iron redox couples as the electroactive species were employed to investigate the transport behavior of vanadium ions in the presence of an electric field. It was shown that the transport of vanadium ions was affected by a combination of different factors, such as the charge/discharge process, current density, and SOC. The electric field accelerated the positive-to-negative and reduced the negative-to-positive transport of vanadium ions in the charging process and affected the vanadium ion transport in the opposite way during discharge. With the increase of current density, the vanadium ion transport rate increases during the charge process but decreases during discharge. At each given current density, the vanadium ion transport rate, in general, demonstrated a trend of decreasing with the increase of SOC.

# (774d) Electrolyte Optimization for Fe/V Redox Flow Battery System

**Bin Li**[1], Wei Wang[1], Zimin Nie[2], Vincent Sprenkle[1], Baowei Chen[3], Xiaoliang Wei[1], Qingtao Luo[1], Guanguang Xia[1], Liyu Li[4] and Zhenguo Yang[4], (1)Pacific Northwest National Laboratory, Richland, WA, (2)Pacific Northwest National Labratory, Richland, WA, (3)Biological Sciences, Pacific Northwest National Laboratory, Richland, WA, (4)UniEnergy Technologies, LLC, Mukilteo, WA

Electrolyte optimization for Fe/V redox flow battery system

Bin Li, Wei Wang, Zimin Nie, Vince Sprenkle, Baowei Chen, Xiaoliang Wei, Qingtao Luo, Guanguang Xia, Liyu Li and Zhenguo Yang

Pacific Northwest National Laboratory

902 Battelle Boulevard, PO Box 999, Richland, WA 99354

**Abstract**

Redox flow batteries (RFBs) research and development have recently enjoyed a renaissance primarily due to their ability to store large amounts of power and energy, up to multi-MW and multi-MWh, respectively, for renewable energy integration or smart-grid deployment.[1] Broad market penetration, however, has not been achieved due to several limitations such as high cost, instability in long-term operation, and narrow operational temperature range, etc. Recently invented Fe/V redox flow battery (IVRFB) has attracted more and more attentions due to its long-term cycling stability and the freedom to utilize membranes other than expensive Nafion.[2] In this study, extensive matrix study was performed on the Fe/V electrolyte in order to determine the relationships of the composition, state of charge (SOC), and temperature stability. As we know, the temperature stability of the RFB systems is of paramount importance for practical use. A RFB system with wider temperature stability will not only enable the RFB to be deployed with less geographical limitation, but also possibly eliminate the active temperature management system reducing parasitic loss. As previously reported, the current Fe/V electrolyte cannot be operated or stored at lower temperatures ($\leq 0^oC$) due to the precipitations ($FeCl_2$). In this research, factors influencing the stability of electrolytes in both positive and negative half-cells with different SOCs at different temperatures from $-10^oC$ to $40^oC$ were systematically studied. Through the optimization of the electrolyte composition, an electrolyte stable in the temperature range from $-5^oC$ to $40^oC$ was identified. Electrolyte stability data and IVRFB flow cell cycling performances using low-cost separators will be presented (Fig.1).

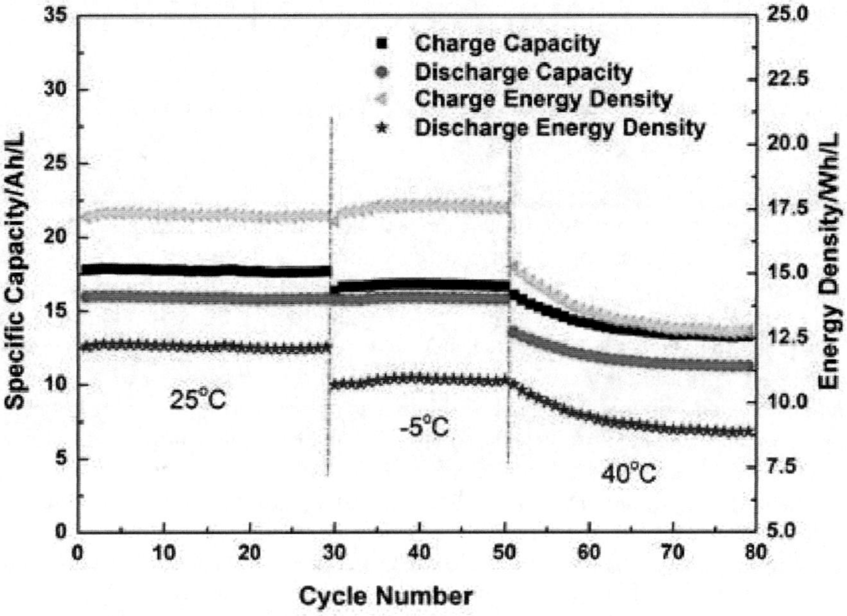


**Fig.1** Cycling performances of optimized Fe/V electrolytes

**References**

(1) Yang, Z.; Zhang, J.; Kintner-Meyer, M. C. W.; Lu, X.; Choi, D.; Lemmon, J. P.; Liu, J. *Chemical Reviews* **2011**, *111*, 3577.

(2) Wang, W.; Kim, S.; Chen, B.; Nie, Z.; Zhang, J.; Xia, G.-G.; Li, L.; Yang, Z. *Energy & Environmental Science* **2011**, *4*, 4068.

# (774e) Electrostatically Assembled Carbon Nanoparticle Electrodes for Vanadium Redox Flow Batteries

Abhinandh Sankar and **Anastasios Angelopoulos**, School of Energy, Environmental, Biological, and Medical Engineering, University of Cincinnati, Cincinnati, OH

Efficient charge and mass transfer at the surface of the electrode is critical to overall redox flow battery (RFB) performance. Fundamental structure-property correlations for RFB electrodes cannot presently be obtained using electrodes produced from carbon felts or carbon nanoparticle slurries and pastes due to the ill-defined distribution of active sites in these materials. For example, a recent study using high surface area multi-wall carbon nanotubes as electrode materials yielded only a two to three-fold increase in peak current density relative to glassy carbon. We here report on the first use of electrostatic layer by layer (LbL) assembly to fabricate RFB electrodes. Since its introduction by Iler over four decades ago, LbL assembly has been repeatedly shown capable of producing well-defined arrangements of nanoparticles and the method is now ubiquitous in scientific research. However, the approach has yet to be applied to produce well-defined electrode structures for RFB electrodes and has only rarely been explored in the general field of electrocatalysis. Key areas of concern include: 1) the impact of the polymer counter-ion layer that is typically required for assembly on ohmic resistance, 2) accessibility of active sites in tightly packed layered structures, and 3) durability. In this investigation, we demonstrate how LbL assembly of graphite nanoplatelets may be used to achieve unprecedented high current densities for the all-vanadium system through fine control over the thickness and morphology of the electrode. In particular, we show how nanoparticle packing density, size, and size distribution in addition to local polymer binder concentration and crystalline order all impact active site concentration and promote reagent accessibility to these sites in a manner that has not been previously recognized. Such basic understanding permits rational optimization of electrode chemistry and structure using carbon-based electrocatalysts as well as detailed interpretation of the promotional effects of inorganic electrode additives. Finally, we demonstrate through electrochemical cycling how these coatings are exceptionally stable relative to electrodes fabricated without LbL assembly.

# (774f) Fractional Degradation Cost for Cycling a Zinc/Bromine Flow Cell Battery

Alex Bistrika[1], Alexandre F.T. Yokochi[1], Michael Antonishen[2], James Davidson[2], Douglas Halamay[2], David Naviaux[3], Jiajia Song[2], Ted Brekken[3] and Annette von Jouanne[3], (1)School of Chemical, Biological and Environmental Engineering, Oregon State University, Corvallis, OR, (2)Electrical Engineering and Computer Science, Oregon State University, Corvallis, OR, (3)School of Electrical Engineering and Computer Science, Oregon State University, Corvallis, OR

It is widely understood that intermittent resources such as solar and wind are dependent on environmental factors outside our ability to control. This allows for the possibility of either power output exceeding or falling short of forecast levels which leads to more frequent cycling of regional hydro generators and greater overall grid instability under the current infrastructure. Quantifying the cost associated with added maintenance attributed to an increased penetration of wind and the physical limitations of a growing dependence on Pumped Hydro Energy Storage (PHES), becomes essential in painting a more complete picture. However, alternative means of Large Scale Energy Storage (LSES), for storing excess power when the outputs exceeds demand and to supplement output power when it falls short of demand, is a widely accepted route for mitigating intermittent power generation.[1] Thus, it is crucial to establish an accurate and relative cost function for the commercially available and most economically viable storage technologies. The most promising electrical energy storage technology, to replace PHES, is flow cell batteries.[2] This begs the question what are the degradation mechanisms in these building size chemical systems and what do we do when they stop working. In this study we report on the key degradation mechanisms that could be expected for standard operation of a zinc/bromine flow cell battery using graphite electrodes and assign a cost function dependent of system operating parameters. In a previous study we reported the oxidation (or charging) reaction as the most damaging to the electrode material, see **Figure 1**. We discuss aging trends and electrode performance by considering surface chemistry as observed by Raman spectroscopy and scanning electron microscopy. Theoretical system performance based on steady state current density measurements is used to establish a normalized operating cost function for a zinc/bromine battery, as can be seen in **Figure 2**.

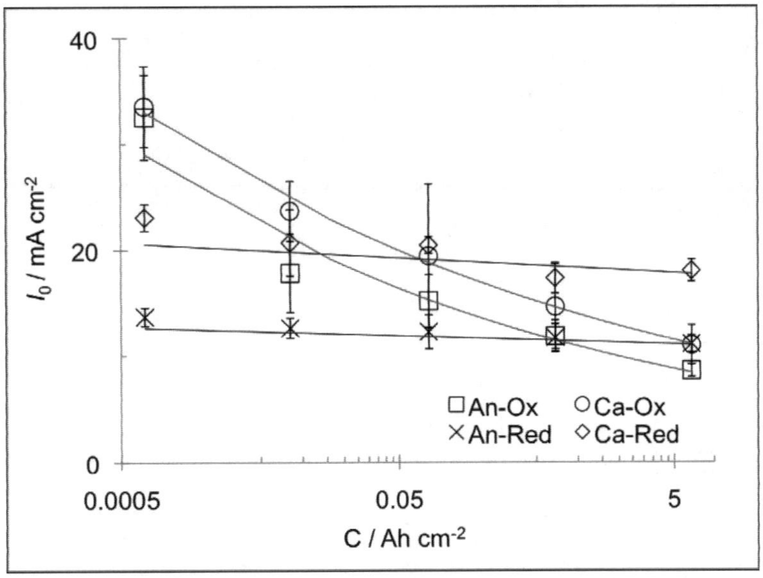


Figure 1 Oxidation and reduction exchange current; measured for both the cathode and anode in 6 [M] NH4Br. The x-axis represents the electrode aging, depicted by total coulombs passed through the electrode per unit surface.

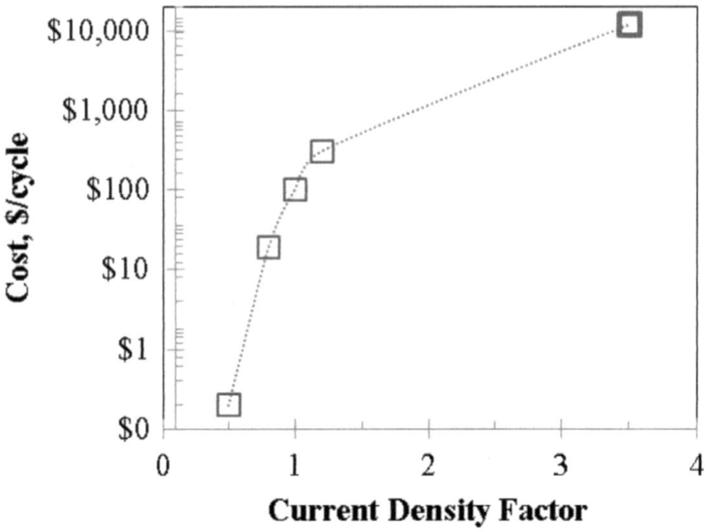


Figure 2 Cost of a 10 % state of charge cycle with respect to a representative operating parameter (current density factor).

**1.** Yang Z, Zhang J, Kintner-Meyer MCW, et al. Electrochemical Energy Storage for Green Grid. *Chemical Reviews.* 2012/04/07 **2011**;111(5):3577-3613.

**2.** Soloveichik GL. Battery Technologies for Large-Scale Stationary Energy Storage. *Annual Review of Chemical and Biomolecular Engineering.* 2012/04/24 **2011**;2(1):503-527.

See more of this Session: Sustainable Electricity: Generation, Transmission, and Storage I
See more of this Group/Topical: Sustainable Engineering Forum

# (774g) Design Mesostructured LiMn2O4 As Low-Cost Cathode Materials for Lithium Ion Batteries

**Jian Zhu**, Chemical Engineering, Wayne State University, Detroit, MI, K. Y. Simon Ng, Department of Chemical Engineering and Materials Science, Wayne State University, Detroit, MI and Da Deng, Chemical Engineering and Materials Science, Wayne State University, Detroit, MI

Lithium ion batteries are the dominant power sources for mobile electronic devices and will play increasingly important role in expanding market of electric vehicles and green energy storage. Spinel $LiMn_2O_4$, alternative to expensive and toxic $LiCoO_2$, as cathode materials in lithium ion batteries are attracting much attention for advantages of low cost, environmental benignness, and abundance. However, the low lithium-ion diffusion coefficient and capacity fading caused by dismutation reaction of $Mn^{3+}$ restrict the applications of $LiMn_2O_4$ at large/industrial scale where bulky batteries are required to have good cycling performance at high current rate with no significant capacity fading. Our strategy is to design and tailor the structure and composition of $LiMn_2O_4$ to improve its conductivity and cycling performance. In this presentation, our latest efforts on the design and facile fabrication of $LiMn_2O_4$ as high-power cathode materials will be reported. The unique structure is tailored and controlled to allow fast lithium ion diffusion while maintaining structure stability. The unique mesostructure is fully characterized by electron microscopy and x-ray diffraction and their electrochemical performance in lithium ion batteries is thoroughly evaluated.

# (791a) Nanomaterials Enabled Lithium Chemistry for Advanced Rechargeable Batteries

**Chao Wang**, Materials Science Division, Argonne National Laboratory, Argonne, IL

Employment of nanomaterials in electrode assembly is promising for the development of advanced lithium-ion batteries. Nanostructured electrodes can potentially achiever higher power and better cyclability than their bulk counterparts, due to the facilitated charge transfer at electrode/electrolyte interface, electronic/ionic transport in the electrodes, and accommodation of strain during lithium insertion/extraction by reduction of material dimensions. Moreover, design of electrodes based on nanomaterials can enable new lithium insertion/extraction reactions that are not possible with bulk materials, and thereby expand our scope of material choices for energy storage. These advantages have lead to substantial literature effort on the synthesis of electrochemically active nanomaterials for battery applications.

Here we report the development of novel manganese oxide nanoparticles (10 - 100 nm) by organic solution synthesis and their solid chemical conversion into lithium manganese oxide ($Li_xMn_2O_4$) spinels for cathode application in lithium ion batteries. We managed to demonstrate that by reducing the material dimensions to nanoscale, charge/discharge up to $0<x<2$ can be achieved which however is not possible for conventional bulk or microscale materials (with only $0<x<1$). Our work shows the promise of expanding the scope of material choices for energy storage by size scaling and the great of potential of improving battery performnace by employing advanced nanomaterials.

# (791b) Dynamics of Pumped Heat Energy Storage Systems

**Hugo S. Caram** and Fan Ni, Chemical Engineering, Lehigh University, Bethlehem, PA

The eventual success and widespread utilization of photovoltaic and wind energy generation will depend on the availability of efficient and economical storage capability. The current technology for long term (diurnal), high efficiency large scale energy storage of electrical energy appears to have been limited to Pumped Hydroelectric Storage (PHS) and Compressed Air Storage (CAES). Both of them are dependent of the availability of geological features such as elevated lakes or empty deep mines for PHS or caverns for the CAES. This features are not available everywhere and other technologies such flow batteries while capable of large scale storage are limited in their power output.

Pumped heat energy storage systems can provide a valuable alternative PHS or CAES. The proposed process uses a Brayton cycle where during the loading cycle heat is pumped to heat up a high pressure thermal regenerator while a low pressure thermal regenerator is cooled down to a low temperature. During the unloading period the gas circulation is reversed. Cold gas from the low pressure regenerator is compressed and heated in the high pressure regenerator to be then expanded to the low pressure vessel. Electrical energy storage capacities in the 30-50kWh/m$^3$ range appear possible and are comparable to lead acid batteries (25-50 kWh/m$^3$).

Analysis and optimization of these systems are complicated by the difficult dynamics of thermal regenerators. A novel matrix algorithm is proposed that allows us to directly calculate cyclic steady state as well as transients. The effects of different design parameters such as pressure ratios, compressor and expander efficiencies, type of gases, solid heat capacities and gas solid heat transfer coefficients are explored. Possible synergistic integration with natural gas combustion combined cycles will be discussed.

# The effective utilization System for Renewable Energy Making use of Regional Characteristic

Masahiro Yamamoto, Tsuguhiko Nakagawa

Graduate school of *Okayama Prefectural University,*

*Kuboki 111, Soja-shi, 719-1197, Okayama, Japan*

## ABSTRACT

The SMART factory project has been developing a process to reduce $CO_2$ emissions through researching and developing a process to make plastic from biomass at the biomass accumulated factory in Maniwa, Okayama.

At the biomass accumulated factory, wood chips have been made from shavings and unused wood parts from sawmills. The biomass factory consists of a building chips plant which makes the wood chips, a building silo which stores chips, and a barkyard which crushes bark. The energy used at the biomass accumulated factory consists of electric power and liquid fossil fuels. Liquid fossil fuels have been used by forklift trucks and others vehicles for wood chips. The $CO_2$ emissions by liquid fossil fuels are accounted for about 43% of the total from the biomass accumulated factory. In order to reduce 25% $CO_2$ emissions, the SMART factory is planned to be supplied by renewable energy from the wood chips plant and photovoltaic power generation.

In this study, in order to achieve this aim, the mathematical simulation model which calculates the energy demand and supply per hour in the SMART factory has been developed. Based on the mathematical simulation model, a system which is able to reduce consumption of liquid fossil fuels and electric power has been designed.

As a result, it has been proved that it is possible to reduce 25% or more $CO_2$ emissions using the system that combines photovoltaic power generation with electric forklift trucks.

## 1. INTRODUCTION

The SMART factory project in Maniwa has been developing a process to reduce $CO_2$ emissions using renewable energy through researching and developing a process to make plastic from biomass at the biomass accumulated factory.

In this study, we have examined the system which substituted fossil fuels to be used as fuel for mobile equipment such as forklift trucks for renewable energy such as photovoltaic power generation (hereafter refered to as "PV") which will be constructed at the biomass accumulated factory.

## 2. OUTLINE OF THE NEW ENERGY SYSTEM

Outline of the new energy system is shown in Fig.1. At the biomass accumulated factory, mobile equipment such as two forklift trucks, shovel loaders and two grapples have been used. These vehicles have previously used liquid fossil fuels. Among this mobile equipments we considered that forklift trucks be substituted for electric forklift trucks and the system that charges the PV electricity to the battery of electric forklift trucks.

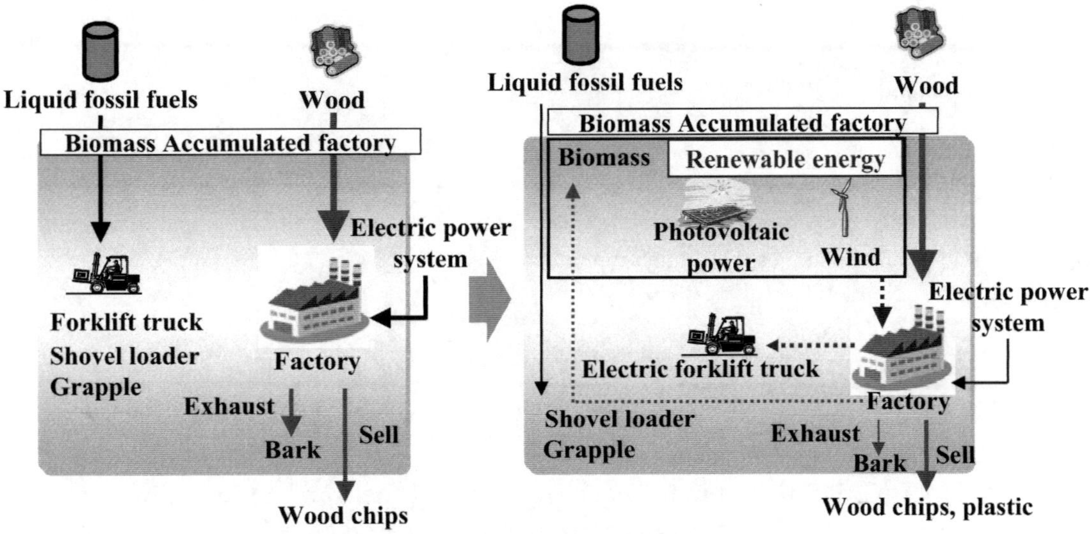


Fig.1. Images of the new energy system

## 3. THE EXAMINATION METHOD

### *3.1 Model Equation*

For each production of wood chips, using equation (1), the mathematical simulation model which is able to calculate the energy demand and supply per hour in a year (8760hr) has been designed.*[1]

$$\sum_{i=1}^{n} \alpha_i = \frac{X_i}{\eta} + E_i - \beta_i \quad \cdots\cdots (1)$$

where $\alpha_i$ : electric power from electric power system [kWh]

$E_i$ : electric power demand of factory [kWh]

$X_i$ : electric power demand of electric forklift truck [kWh]

$\beta_i$ : electric power generation of PV [kWh]

$\eta$ : Charging&discharging efficiency (=0.85)

### *3.2 Production of wood chips and energy consumption*

The relationship of monthly production of wood chips and electric power demand, liquid fossil fuels consumption of forklift truck at the biomass accumulated factory is shown in Fig.2 and Fig.3.

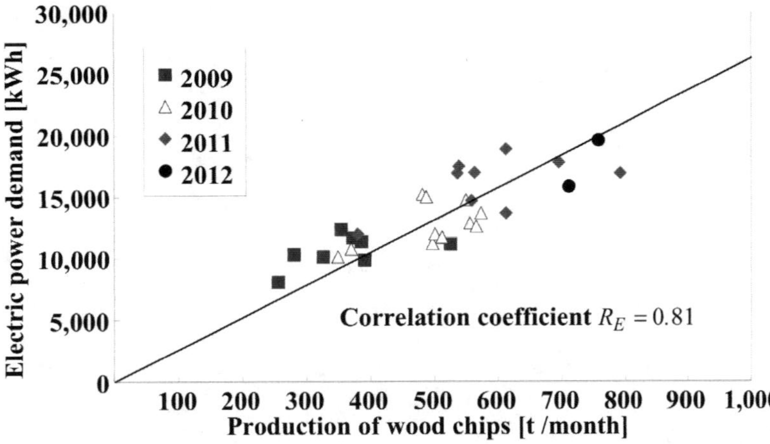


Fig.2. Relationship between electric power demand and production of wood chips

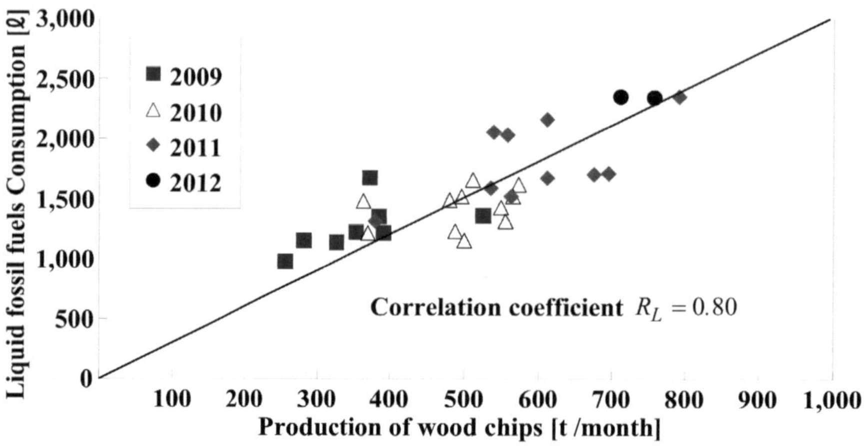


Fig.3. Relationship between liquid fossil fuels consumption of forklift trucks and production of wood chips

In Fig.2, it can be seen that there is a correlation between electric power demand and production of wood chips. For least-squares method,

$$y = 26.3x \quad \cdots\cdots\cdots \quad (2)$$

From equation (2), it can be estimated electric power demand from production of wood chips. On the other hand, in Fig.3, similarly, it can be obtained by equation (3).

$$Z = 0.9x \quad \cdots\cdots\cdots \quad (3)$$

where x : Production of wood chips [ t/month ]
y : electric power demand [kWh]
Z : liquid fossil fuels consumption of forklift trucks [ ℓ ]

Using equations (2) and (3) in this new energy system the optimal operation performance can be calculated depending on the production level of wood chips.

### *3.3 Result of examination*

Regarding production of wood chips at 400[t/month] and 700[t/month], the relationship of $CO_2$ emissions and area of PV at the biomass accumulated factory is shown in Fig.4.

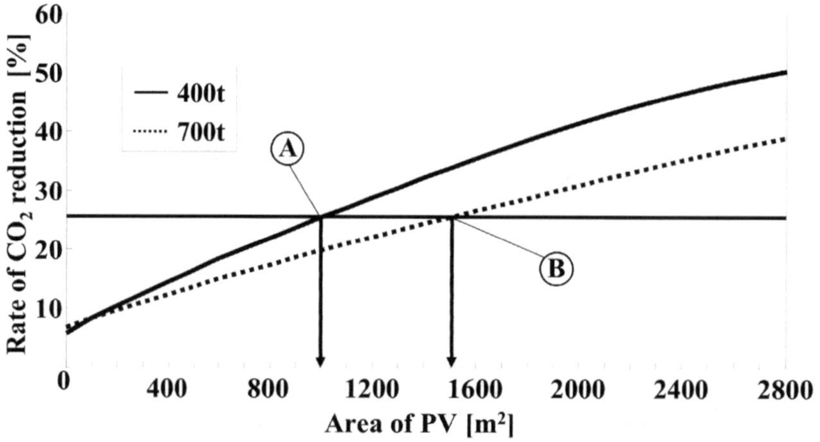


Fig.4.Relationship area of PV and rate of $CO_2$ reduction per production of wood chips

In Fig.4Ⓐ,Ⓑ the breakdown in the case of reducing 25% $CO_2$ emissions is shown Fig.5 Ⓐ,Ⓑ. In Fig.4, PV is of low electric power generation efficiency. But, it is able to be installed on a sloping site e.g. the factory roofs. Also, it is amorphous that installation costs are low. Maximum output per unit area is 69.1[W/m$^2$].

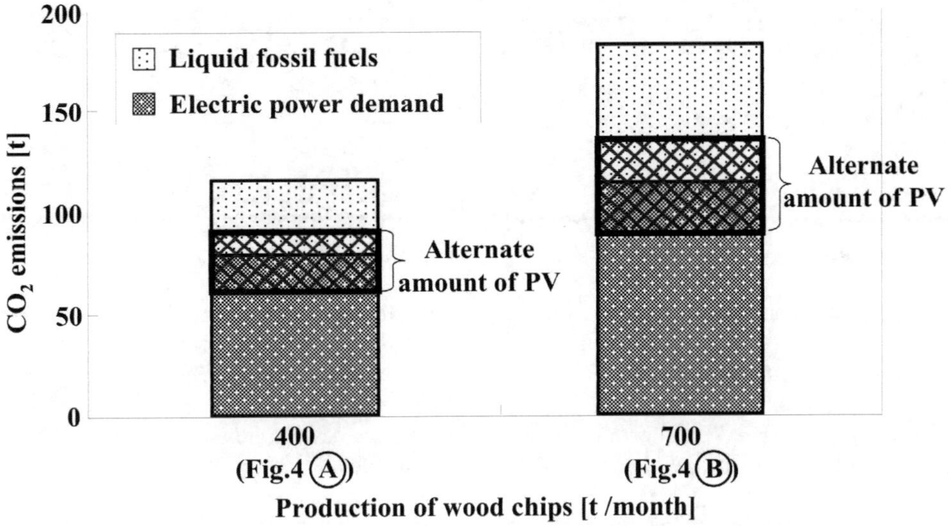


Fig.5. The breakdown in the case of reducing 25% $CO_2$ emissions

As a result, when production of wood chips is 700[t/month], it has been proved that the PV installation area should be greater than 1500m$^2$ to reduce 25% or more $CO_2$ emissions at the biomass accumulated factory.

## 4. CONCLUSION

Depending on the rate of Production of wood chips, it has been proved that it is possible to estimate liquid fossil fuels and electric power demand and calculate the design area of PV depending on the target of $CO_2$ emissions reduction.

In this study, we are subsidized by the funds form the Overall Promotion of Ministry of Education, Culture, Sports, Science and Technology. I am deeply grateful to them for their supports.

***Reference***

1. Shimohara, K., Y. Notoji, T. Nakagawa "Distributed Energy System Combination with Energy Conservation, PV and EV", *Kagaku kogaku Ronbunshu* Vol.38 No.4, pp.255–262, Japan (2012)

# (791d) Thermochemical Cycle of a Mixed Metal Oxide for Augmentation of Thermal Energy Storage in Solid Particles

**Brian D. Ehrhart**[1,2], Nathan P. Siegel[3], Eric N. Coker[2] and Alan W. Weimer[1], (1)Chemical and Biological Engineering, University of Colorado at Boulder, Boulder, CO, (2)Sandia National Laboratories, Albuquerque, NM, (3)Mechanical Engineering, Bucknell University, Lewisburg, PA

Solid particle heat transfer and storage media have been shown to be able to operate at temperatures >1000°C in concentrated solar power (CSP) applications, much higher than the operational limit of 600°C for currently state-of-the-art molten nitrate salt heat transfer fluid. Solid particles can be endothermically reduced by direct exposure to concentrated solar energy, thus absorbing and storing thermal energy beyond that possible with sensible heating alone. The particles can then be oxidized exothermically at a later time, releasing the stored chemical heat and effectively augmenting the thermal energy storage capacity of the solid particles. A mixed metal oxide spinel material that reduces in the temperature range of interest (1000°C-1200°C) has been examined for applicability to this thermochemical energy storage concept. A description of this application, prospective materials, and details of the thermochemical cycle are presented. The heats of reduction and oxidation for the thermochemical cycle have been determined for various operating conditions to evaluate the amount of thermal energy that may be stored. Additionally, the extent of reduction possible at various operating conditions has been measured.

# (791e) Carbohydrate Could Be the Best Electricity Storage Compound and a High-Density Hydrogen Carrier

**Y.-H. Percival Zhang**, Biological Systems Engineering, Virginia Polytechnic Institute and State University, Blacksburg, VA and Chun You, Biological Systems Engineering, Virginia Polytechnic Institute and State University

Electricity is a high-quality secondary energy, which can be generated from renewable primary energies and fossil fuels. Carbon-neutral electricity generation is becoming more and more important, but most of them, such as wind and solar electricity, suffer from large output variations. Therefore, it is vital to store electricity in the form of chemical compounds. Among numerous potential chemical compounds, such as hydrogen, isobutanol, methanol and hydrocarbons, we suggest that carbohydrates (e.g., starch, cellulose, glucose) could be the best electricity storage compound with a combustion energy density of 17 MJ/kg and an electricity output density of >3,000 Wh/kg [1]. We designed a novel biological CO2 fixation pathway that can fix CO2 by using hydrogen or electricity for producing carbohydrate mediated by cell-free cascade enzymes [1]. In this talk, we will present our efforts in developing building blocks and building modules for this ambitious project and conduct proof-of-concept experiments. Through the above bioconversion plus our previous achievements such as theoretical conversion of carbohydrate to hydrogen or electricity [2] and low-cost enzymatic conversion of cellulose to starch (unpublished), the electricity-carbohydrate-hydrogen cycle would not only supplement current and future primary energy utilization systems for facilitating electricity and hydrogen storage and enhancing secondary energy conversion efficiencies, but also address such sustainability challenges as transportation fuel production, $CO_2$ utilization, fresh water conservation, food and feed production, and maintenance of a small closed ecosystem in emergency situations [1]. In a word, carbohydrate would become a new oil in the future.

1. Zhang Y-HP, Huang W-D: **Constructing the electricity-carbohydrate-hydrogen cycle for a sustainability revolution**. *Trends Biotechnol.* 2012, **Epub. DOI: 10.1016/j.tibtech.2012.02.006**.

2. Zhang Y-HP, Evans BR, Mielenz JR, Hopkins RC, Adams MWW: **High-yield hydrogen production from starch and water by a synthetic enzymatic pathway**. *PLoS One* 2007, **2**:e456.

# (791f) Energy Storage Design for Electric Transmission Systems

**Donald J. Chmielewski**, Illinois Institute of Technology, Chicago, IL

It is widely recognized that a major concern with the installation of renewable energy systems is the fact that wind and solar sources are non-dispatchable. That is, the power produced from renewable systems is dependent on environmental conditions and is likely uncorrelated with the power demand from load centers. In contrast, fossil based sources typically have the ability to track demand, but depending on the technology possess limits with respect to response time. Thus, many have advocated the use of massive energy storage systems to capture energy produced from renewable sources and then dispatch as load centers demand. An additional concern is the fact that renewable sources tend to be located far from the load centers (again in contrast to most fossil sources which are typically placed near demand). This combination of spatial and temporal mismatch with load center demand has caused many to worry about the stability of the electric power grid. Clearly, wide swings in power throughput over great distances will cause the weakest parts of the grid to fail.

To address both the spatial and temporal issues surrounding renewable power generation we have developed a scheme aimed at analyzing the capabilities of a regional power generation and distribution system. In particular, the scheme proposes a supervisory control law aimed at utilizing a mix of renewable and fossil sources along with energy storage systems to meet power load demands. This proposed control law takes into consideration not only the power output capacity of each source but also the rate at which power output can change as well as the storage capacity and efficiency losses associated with the storage units. In addition, the proposed scheme incorporates the maximum and minimum throughput limits on the distribution lines. If time allows, we will present an extension of the scheme aimed at sizing and placing energy storage systems as well as locating new power distribution lines.

# AUTHOR INDEX

# AUTHOR INDEX

# AUTHOR INDEX

# AUTHOR INDEX